Sliding Base Structures

Hong-Song Hu

Sliding Base Structures

Analysis and Design

 Springer

Hong-Song Hu
College of Civil Engineering
Huaqiao University
Xiamen, China

ISBN 978-981-99-5106-2   ISBN 978-981-99-5107-9 (eBook)
https://doi.org/10.1007/978-981-99-5107-9

This Springer imprint is published by the registered company Springer Nature Singapore Pte Ltd.
The registered company address is: 152 Beach Road, #21-01/04 Gateway East, Singapore 189721, Singapore

Preface

After getting my Ph.D. in 2014, I came to Kyoto University for postdoctoral research. My host, Prof. Masayoshi Nakashima, invited me to participate in a research project on free-standing structures. The free-standing structure has the configuration that the superstructure is disconnected from the foundation, and the interface between the superstructure base and the foundation is lubricated with graphite to reduce the friction coefficient. This idea is very close to sliding isolation. Professor Nakashima intended to use this technology in structural retrofitting, so he used a new name.

After performing a preliminary literature review, I found that scientific research on sliding base (SB) structures (I used this name in my later papers, and in some early papers, "sliding structures" is commonly used) started as early as the 1980s. Both Westermo and Udwadia and Mostaghel et al. published a paper on the response of SB structures subjected to harmonic excitation in 1983. Afterward, other researchers (e.g., Jangid R. S. at the Indian Institute of Technology) also conducted several analytical studies on SB structures. However, there still existed some research gaps in this topic prior to 2014. For instance, the theoretical solutions for the sliding-sliding case had not been derived; the former studies only used a small number of ground motion records, and simplified equations for estimating the peak superstructure response and the peak sliding displacement had not been developed.

During my one year and two months stay in Japan, I completed a study on the dynamic responses of two-degree-of-freedom SB systems subjected to harmonic excitations. Meanwhile, I took part in the shaking table tests of free-standing structures, which was a unique research experience for me.

After completing my postdoctoral fellowship, I joined Huaqiao University as a faculty. My research focus returned to the steel-concrete composite structures, my main research interest. However, I still spare some time to continue the left work on SB structures. From 2016 to 2022, I, working with two of my graduate students, Fan Lin and Li-Wen Xu, completed studies on the peak responses and design methods of SB structures subjected to three-component earthquake excitations. My colleague, Prof. Yi-Chao Gao, helped me develop the program for response history analyses of SB structures subjected to three-component excitations. I am very grateful for his help.

I wrote this book based on my research on SB structures in the past eight years. I hope this book's publication can guide the design of SB structures and promote practical applications of SB structures.

Xiamen, China   Hong-Song Hu

Contents

Chapter 1
Introduction

1.1 Fundamentals of Sliding Base (SB) Structures

Base isolation is an effective approach for reducing damages to structures and their contents under severe earthquake excitations. It is generally implemented by using special isolators, such as laminated rubber bearings (e.g., Kelly, 1986; Kikuchi & Aiken, 1997; Skinner et al., 1993; Yamamoto et al., 2009) and friction pendulum (FP) bearings (e.g., Becker & Mahin, 2013; Castaldo & Tubaldi, 2015; Mokha et al., 1991; Roussis & Constantinou, 2006). The properties of these isolators can be elaborately designed to achieve a certain structural performance. However, they are expensive and require high construction techniques. Therefore, when the cost is a major concern, base isolation using isolators may not be an appropriate choice.

Adopting a sliding interface between the base of the superstructure and the foundation (Fig. 1.1) can also reduce the seismic response of the superstructure. The mechanism is very simple: as the friction force between the superstructure and the foundation has an upper limit, the seismic force transmitted to the superstructure is limited. Structures that adopt this type of isolation technique are called sliding base (SB) structures in this book. Since the implementation of SB structures is simple and cost effective, they are applicable to low-rise buildings in rural areas. Actually, SB structures have been used in some low-rise masonry buildings (Li, 1984; Zhou, 1997).

1.2 Practical Implementations of SB Structures

For the past four decades, several materials have been investigated regarding their potential use along the sliding interface of SB structures. Qamaruddin et al. (1986) conducted shaking table tests on sliding brick building models with different sliding layer materials, namely, graphite powder, dry sand, and wet sand, and obtained friction coefficients of 0.25, 0.34 and 0.41, respectively, for the corresponding interfaces.

© The Author(s) 2023
H.-S Hu, *Sliding Base Structures*,
https://doi.org/10.1007/978-981-99-5107-9_1

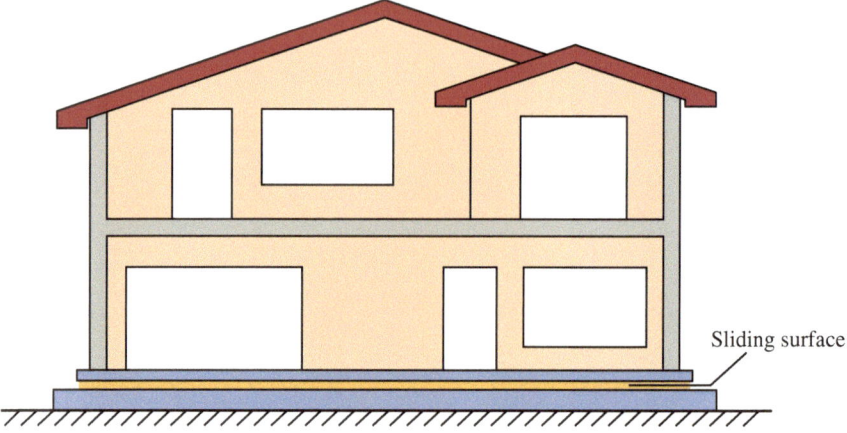

Sliding surface

Fig. 1.1 Schematic plot of a building adopting SB technique

Tehrani and Hasani (1996) conducted experimental studies on adobe buildings with dune sand and lightweight expanded clay as sliding layers; the friction coefficients were reported as 0.25 for dune sand and 0.2–0.3 for lightweight expanded clay. Barbagallo et al. (2017) tested a steel-mortar interface lubricated by graphite powder; the static and dynamic friction coefficients were close to 0.19 and 0.16, respectively, and they were independent of both the sliding velocity and the superstructure properties.

Polymer materials are also suitable choices for the sliding interface. Yegian et al. (2004) investigated the frictional characteristics of four synthetic interfaces [namely, geotextile-high density polyethylene (HDPE), polypropylene (PTFE)-PTFE, ultrahigh molecular weight polyethylene (UHMWPE)-UHMWPE, and geotextile-UHMWPE] as potential candidates for sliding isolation through cyclic and shaking table tests. It was determined that the geotextile-UHMWPE interface was suitable for sliding isolation applications because the friction coefficient of this interface is insensitive to large variations in the sliding velocity and normal stress; as a result, this interface can easily be introduced into engineering design. The obtained static and dynamic friction coefficients of the geotextile-UHMWPE interface were approximately 0.11 and 0.08, respectively. Nanda et al. (2012, 2015) conducted experimental studies on four sliding interfaces with green marble against HDPE, green marble, geosynthetics and rubber sheeting, respectively. The static friction coefficients were found to be independent of the normal stress, and the dynamic friction coefficients were insensitive to variations in the sliding velocity. Moreover, the observed dynamic friction coefficients of the four investigated interfaces ranged from 0.07 to 0.18, and the relative differences between the static and dynamic friction coefficients were all below 15%. Jampole et al. (2016) adopted sliding isolation bearings consisting of HDPE sliders and galvanized steel surfaces to seismically isolate light-frame residential houses; shaking table tests showed that the friction coefficient of this sliding interface was nearly 0.18 with a slight variation between the stick and sliding phases.

The aforementioned sliding interfaces were basically insensitive to the variations in the sliding velocity and pressure. Therefore, the Coulomb friction model can be used to model the behavior of the sliding interface of an SB structure.

1.3 Review of Analytical Studies on SB Structures

1.3.1 Studies on 2DOF SB Systems Subjected to Harmonic Ground Motions

The simplest model for an SB structure contains two masses, one for the superstructure and the other for the sliding base; thus, this model is referred to as a 2-degree-of-freedom (2DOF) SB system. Westermo and Udwadia (1983) and Mostaghel et al. (1983) first studied the dynamic responses of 2DOF SB systems under harmonic excitations, and both groups developed the governing equations of motion and numerical implementations needed to perform response history analyses of such systems. Westermo and Udwadia (1983) pointed out that the response of this system under harmonic excitations converged rapidly to a periodic response after several cycles. Three different periodic responses were observed, namely stick-stick, stick-sliding, and sliding-sliding cases, depending on the amplitude of the input accelerations and the structural characteristics of the system. They also derived the explicit equations for the condition of the initiation of the stick-sliding case. Mostaghel et al. (1983) conducted parametric studies for the critical responses of the 2DOF SB system under harmonic ground motions. Iura et al. (1992) followed the work of Westermo and Udwadia (1983), and Mostaghel et al. (1983). They derived the analytical expressions for the condition of the initiation of the sliding-sliding case. Therefore, combined with the work of Westermo and Udwadia (1983), the explicit equations for the occurrence conditions of three periodic response cases were obtained. More recently, Hu and Nakashima (2017) conducted a comprehensive parametric study on the maximum responses of 2DOF SB systems under harmonic ground motions and derived a theoretical solution for the response corresponding to the sliding-sliding case.

1.3.2 Studies on SB Structures Under Earthquake Excitation

Mostaghel and Tanbakuchi (1983) studied the seismic responses of 2DOF SB systems subjected to the N-S component of the El-Centro record from the 1940 Imperial Valley earthquake and the S86E component of the Olympia record from the 1949 Western Washington earthquake. Response spectra of the absolute acceleration and sliding displacement were developed. It was found that the SB isolation can effectively control the level of the superstructure response. Qamaruddin et al. (1986b) conducted a study similar to that of Mostaghel and Tanbakuchi (1983), with emphasis on using the SB system in masonry buildings. The N-S component of the El Centro record and the longitudinal component of the Koyna record were considered. The findings of their study were similar to those of Mostaghel and Tanbakuchi (1983). Yang et al. (1990) and Vafai et al. (2001) studied the responses of multiple-degree-of-freedom (MDOF) SB structures. Their focus was to develop efficient numerical methods for response history analyses, whereas the response characteristics of these structures were not sufficiently addressed.

Jangid (1996a) compared the responses of single-story SB structures subjected to two components and a single component of the El-Centro record; the numerical results indicated that the former increased the sliding displacement and reduced the absolute acceleration of the superstructure in comparison with the latter. Shakib and Fuladgar (2003a) adopted the same model as Jangid (1996a) but included the vertical component in the ground motion input, and three ground motion records were considered, namely, the El-Centro record, the Tabas record from the 1978 Tabas earthquake, and the Renaldi record from the 1994 Northridge earthquake. The effect of the vertical component was highly dependent on the superstructure period and the input ground motions; additionally, the responses of low-period structures could be strongly affected by the vertical component of the ground motion, while this influence was insignificant when the superstructure period exceeded 0.7 s. Jangid (1996b) and Shakib and Fuladgar (2003b) also studied the responses of asymmetric single-story SB structures under multidimensional inputs. Both studies indicated that the bidirectional interaction between the frictional resistance at the sliding interface and the vertical component of the earthquake excitation could significantly affect the responses of torsionally coupled systems with SB isolation. More recently, Hu et al. (2020, 2022) investigated the peak superstructure response and peak sliding displacement of SB structures subjected to three-component earthquake excitation using a large number of ground motion records. The influence of various structural and ground motion characteristics on these two response quantities was comprehensively studied, and simplified design equations were also developed.

Chapter 2
Responses of 2DOF Sliding Base Systems Under Harmonic Ground Motions

2.1 Equations of Motion

For a 2DOF SB system shown in Fig. 2.1, the dynamic equilibrium equations can be written as

$$\begin{cases} m\left(\ddot{u}_g + \ddot{u}_s + \ddot{u}_r\right) + c\dot{u}_r + ku_r = 0 \\ m_b\left(\ddot{u}_g + \ddot{u}_s\right) + m\left(\ddot{u}_g + \ddot{u}_s + \ddot{u}_r\right) = f \end{cases} \tag{2.1}$$

in which m, k, and c refer to the top mass, lateral stiffness, and viscous damping coefficient of the superstructure, respectively; m_b is the mass of the sliding base; $u_g(t)$, $u_s(t)$, and $u_r(t)$ are the ground displacement, sliding displacement, and relative displacement between the top mass and sliding base, respectively; $\dot{u}_g$, $\dot{u}_s$, and $\dot{u}_r$ are the corresponding velocities; $\ddot{u}_g$, $\ddot{u}_s$, and $\ddot{u}_r$ are the corresponding; and f is the friction force between the sliding base and foundation. The first equation of Eq. (2.1) denotes the dynamic equilibrium of the top mass, while the second equation is the dynamic equilibrium of the entire system.

The SB system can display two kinds of phases in its response history: the stick phase and the sliding phase. For the stick phases, the sliding acceleration, $\ddot{u}_s$, is equal to 0, and the sliding friction force is greater than the friction force, f; therefore, Eq. (2.1) leads to

$$\begin{cases} m\ddot{u}_r + c\dot{u}_r + ku_r = -m\ddot{u}_g \\ \alpha\ddot{u}_r + \ddot{u}_g = f/(m + m_b) \\ |f| < (m + m_b)\mu g \end{cases} \tag{2.2}$$

where

$$\alpha = \frac{m}{m + m_b} \tag{2.3}$$

© The Author(s) 2023
H.-S Hu, *Sliding Base Structures*,
https://doi.org/10.1007/978-981-99-5107-9_2

Fig. 2.1 A 2DOF sliding
base system

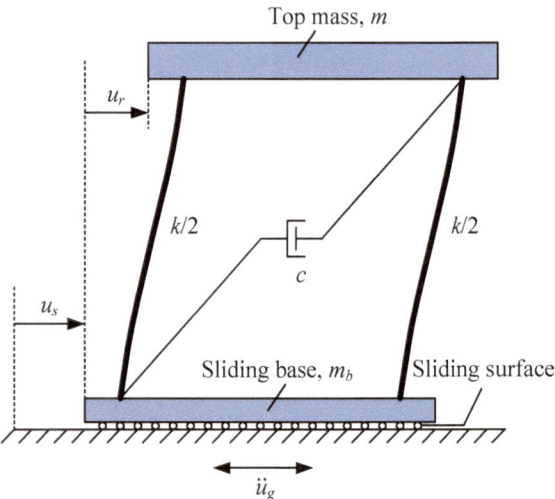

is the mass ratio; μ is the friction coefficient; and g is the gravity acceleration. The first equation of Eq. (2.2) can be written as

$$\ddot{u}_r + 2\xi\omega\dot{u}_r + \omega^2 u_r = -\ddot{u}_g \tag{2.4}$$

where

$$\omega = \sqrt{\frac{k}{m}}, \quad \xi = \frac{c}{2\omega m} \tag{2.5}$$

ω and ξ are the natural frequency and damping ratio of the corresponding fixed base structure, respectively. Equation (2.4) is the differential equation governing the relative displacement of a single-degree-of-freedom (SDOF) system under ground acceleration, $\ddot{u}_g$.

According to the second and third equations of Eq. (2.2), we obtain

$$\left| \alpha\ddot{u}_r + \ddot{u}_g \right| < \mu g \tag{2.6}$$

This is the precondition for the stick phases. Sliding occurs when it is no longer satisfied, and

$$f = \delta(m + m_b)\mu g \tag{2.7}$$

in which δ represents the direction of the friction force, and $|\delta| = 1$. While transitioning from the stick phase to the sliding phase, the direction of the friction force remains unchanged. Thus, it can be inferred from the second equation of Eq. (2.2) that δ and $\alpha\ddot{u}_r + \ddot{u}_g$ have the same sign at the moment before sliding. By substituting

Eq. (2.7) into the second equation of Eq. (2.1), we obtain

$$\ddot{u}_s = \delta\mu g - \alpha\ddot{u}_r - \ddot{u}_g \tag{2.8}$$

Substituting Eq. (2.8) into the first equation of Eq. (2.1) to obtain

$$(1 - \alpha)m\ddot{u}_r + c\dot{u}_r + ku_r = -\delta\mu mg \tag{2.9}$$

Dividing Eq. (2.9) by $(1 - \alpha)m$ to obtain

$$\ddot{u}_r + 2\xi_1\omega_1\dot{u}_r + \omega_1^2 u_r = -\frac{\delta\mu g}{1 - \alpha} \tag{2.10}$$

where

$$\omega_1 = \frac{\omega}{\sqrt{1-\alpha}}, \quad \xi_1 = \frac{\xi}{\sqrt{1-\alpha}} \tag{2.11}$$

Equation (2.10) is the differential equation of a SDOF system with the natural frequency of ω_1 and damping ratio of ξ_1 under a step force. The static displacement corresponding to the step force is

$$u_{st} = \frac{-\delta\mu g}{\omega_1^2(1-\alpha)} = \frac{-\delta\mu g}{\omega^2} \tag{2.12}$$

Thus, in the sliding phases, a new natural frequency, ω_1, and a new damping ratio, ξ_1, of the relative displacement vibration are determined, both of which are linked to the mass ratio, α. The solution of Eq. (2.10) is

$$\begin{pmatrix} u_r(t) \\ \dot{u}_r(t) \end{pmatrix} = \mathbf{A}(\tau)\begin{pmatrix} u_r(t_i) \\ \dot{u}_r(t_i) \end{pmatrix} + \mathbf{b}(\tau) \tag{2.13}$$

in which t is the global time, t_i is the moment when sliding starts, $\tau = t - t_i$, and

$$\mathbf{A}(\tau) = e^{-\xi_1\omega_1\tau}$$

$$\begin{bmatrix} \cos(\omega_{1d}\tau) + \xi_1/\sqrt{1-\xi_1^2}\sin(\omega_{1d}\tau) & \frac{\sin(\omega_{1d}\tau)}{\omega_1\sqrt{1-\xi_1^2}} \\ -\frac{\omega_1\sin(\omega_{1d}\tau)}{\sqrt{1-\xi_1^2}} & \cos(\omega_{1d}\tau) - \xi_1/\sqrt{1-\xi_1^2}\sin(\omega_{1d}\tau) \end{bmatrix} \tag{2.14}$$

$$\mathbf{b}(\tau) = u_{st}\begin{bmatrix} 1 - e^{-\xi_1\omega_1\tau}\left(\xi_1/\sqrt{1-\xi_1^2}\sin(\omega_{1d}\tau) + \cos(\omega_{1d}\tau)\right) \\ e^{-\xi_1\omega_1\tau}\frac{\omega_1\sin(\omega_{1d}\tau)}{\sqrt{1-\xi_1^2}} \end{bmatrix} \tag{2.15}$$

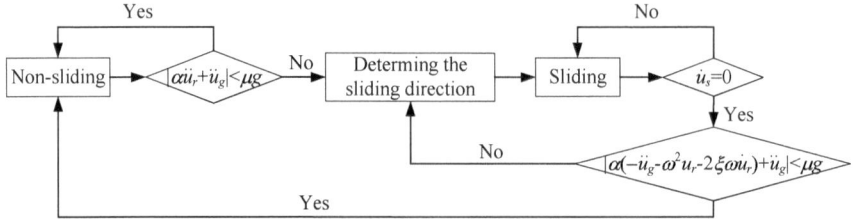

Fig. 2.2 Flow chart for calculating the response history of the 2DOF sliding base system

$$\omega_{1d} = \omega_1\sqrt{1 - \xi_1^2} \tag{2.16}$$

The first term in the right-hand side of Eq. (2.13) corresponds to free vibration caused by the initial condition, while the second term represents the vibration caused by the step force.

At the onset of sliding, the velocity of sliding equals to 0, i.e., $\dot{u}_s(t_i) = 0$. Therefore, integrating Eq. (2.8) leads to

$$\dot{u}_s(t) = \delta\mu g(t - t_i) - \alpha(\dot{u}_r(t) - \dot{u}_r(t_i)) - \left(\dot{u}_g(t) - \dot{u}_g(t_i)\right) \tag{2.17}$$

this round Sliding stops once $\dot{u}_s(t)$ returns to 0 again. The sliding of the structure can either stop or persist, based on whether Eq. (2.6) is satisfied or not. For this check of Eq. (2.6), Eq. (2.4) should be used to determine the relative acceleration $\ddot{u}_r$, under the assumption that sliding stops. The sliding displacement, $u_s(t)$, can be obtained by integrating Eq. (2.17):

$$u_s(t) = u_s(t_i) + \frac{1}{2}\delta\mu g(t - t_i)^2 - \alpha(u_r(t) - u_r(t_i) - \dot{u}_r(t_i)(t - t_i))$$
$$- \left(u_g(t) - u_g(t_i) - \dot{u}_g(t_i)(t - t_i)\right) \tag{2.18}$$

Figure 2.2 summarizes the process of computing the response history.

2.2 Typical Response Histories

In the following analyses, the ground acceleration is taken as a sinusoidal wave if not specified:

$$\ddot{u}_g = a_g\sin(\omega_g t) \tag{2.19}$$

where a_g and ω_g are the amplitude and frequency of the sinusoidal wave, respectively.

Figure 2.3 shows the responses of a sliding base system with $\mu = 0.2$, $\alpha = 0.5$, $T = 2\pi/\omega = 0.5$ s, $\xi = 5\%$, $T_g = 2\pi/\omega_g = 1$ s, and $a_g = 0.4g$, in which T represents the natural period of the superstructure, and T_g indicates the period of the ground acceleration. As shown in Fig. 2.3, the responses of the system converge to steady periodic responses with the same period as the ground acceleration after several cycles. The sliding base system pauses briefly before sliding in the opposite direction in this instance. This is referred to as the stick-sliding case. After the amplitude of the ground acceleration exceeds a certain value, the sliding base system will continue to slide incessantly during the steady periodic state, as shown in Fig. 2.4, where the ground acceleration amplitude, a_g, is increased to $1.2g$. This is referred to as the sliding-sliding case. Sliding will not occur if the ground acceleration is sufficiently low, which is known as stick-stick case. The sliding base structure and the fixed base structure have no difference in this case. The following analyses focus on the steady periodic responses.

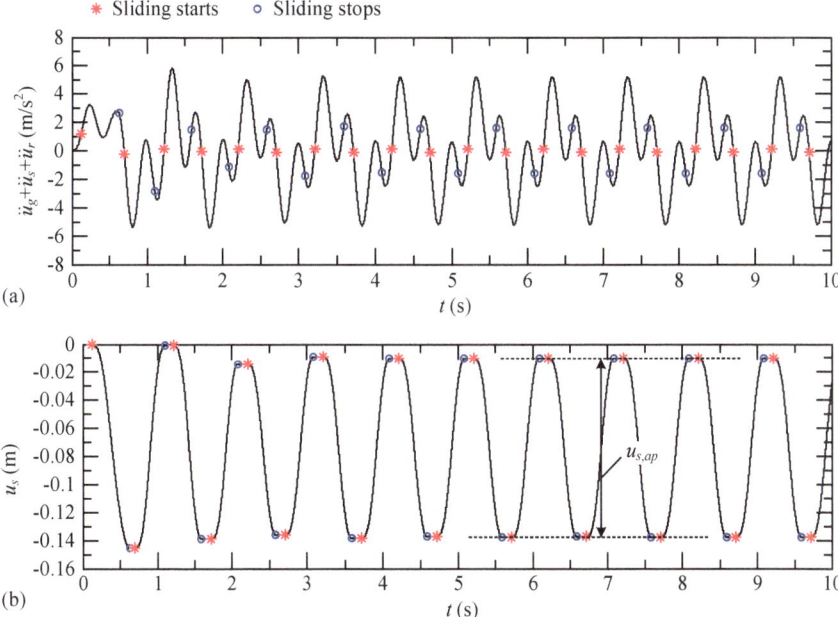

Fig. 2.3 Responses of the sliding base system for $\mu = 0.2$, $\alpha = 0.5$, $T = 2\pi/\omega = 0.5$ s, $\xi = 5\%$, $T_g = 2\pi/\omega_g = 1$ s, $a_g = 0.4g$: **a** absolute acceleration of the top mass; and **b** sliding displacement

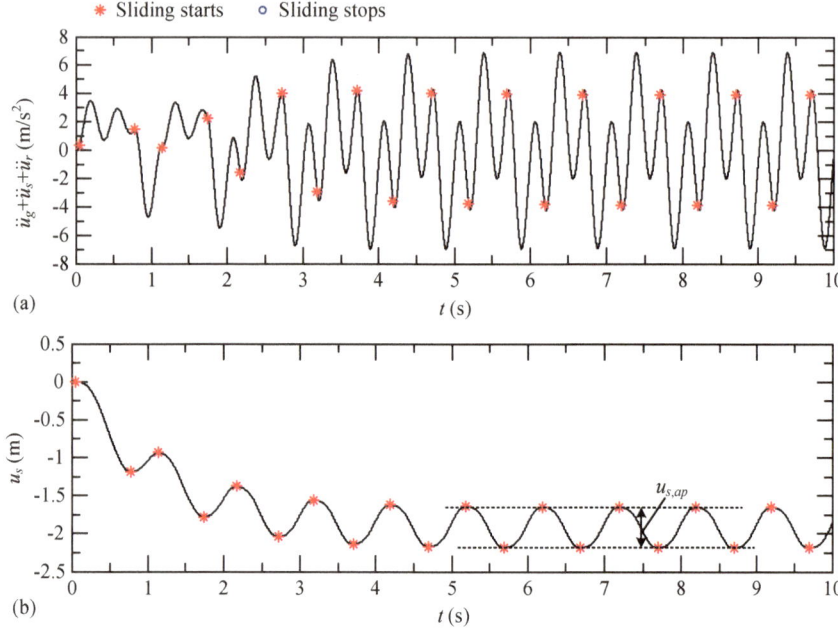

Fig. 2.4 Responses of the sliding base system for $\mu = 0.2$, $\alpha = 0.5$, $T = 2\pi/\omega = 0.5$ s, $\xi = 5\%$, $T_g = 2\pi/\omega_g = 1$ s, $a_g = 1.2g$: **a** absolute acceleration of the top mass; and **b** sliding displacement

2.3   Occurrence Conditions of the Three Types of Periodic Responses

2.3.1   Boundaries Between the Stick-Stick and Stick-Sliding Cases

For the stick-stick case, the steady response of the relative displacement is

$$u_r(t) = -\frac{a_g}{\omega^2} R_d \sin\left(\omega_g t - \phi\right) \tag{2.20}$$

$$\phi = \tan^{-1} \frac{2\xi\left(\omega_g/\omega\right)}{1 - \left(\omega_g/\omega\right)^2} \tag{2.21}$$

where ϕ is the phase angle, which defines the time by which the response lags behind the input ground motion. Substituting Eqs. (2.19) and (2.20) into the precondition for the stick phases (i.e., Eq. 2.6) gives

$$\left|\alpha R_a \sin\left(\omega_g t - \phi\right) + \sin\left(\omega_g t\right)\right| < \mu g/a_g \tag{2.22}$$

where

$$R_a = (\omega_g/\omega)^2 R_d = \frac{(\omega_g/\omega)^2}{\sqrt{\left[1 - (\omega_g/\omega)^2\right]^2 + \left[2\xi(\omega_g/\omega)\right]^2}} \qquad (2.23)$$

is the acceleration response factor for the fixed base structures. Equation (2.22) can be rewritten as

$$\left|(\alpha R_a \cos\phi + 1)\sin(\omega_g t) - \alpha R_a \sin\phi \cos(\omega_g t)\right| < \mu g/a_g \qquad (2.24)$$

Equation (2.24) is always satisfied for the stick-stick cases, so the maximum value of the left-hand side term should be smaller than $\mu g/a_g$; thus, we have

$$\mu g/a_g > \sqrt{(\alpha R_a)^2 + 2(\alpha R_a)\cos\phi + 1} \qquad (2.25)$$

Therefore, the boundaries between the stick-stick and stick-sliding cases are

$$\frac{a_g}{\mu g} = \frac{1}{\sqrt{(\alpha R_a)^2 + 2(\alpha R_a)\cos\phi + 1}} \qquad (2.26)$$

The variations of the boundaries between the stick-stick and stick-sliding cases for different ω_g/ω and α are shown in Fig. 2.5.

Fig. 2.5 Boundaries between the stick-stick and stick-sliding cases ($\xi = 5\%$)

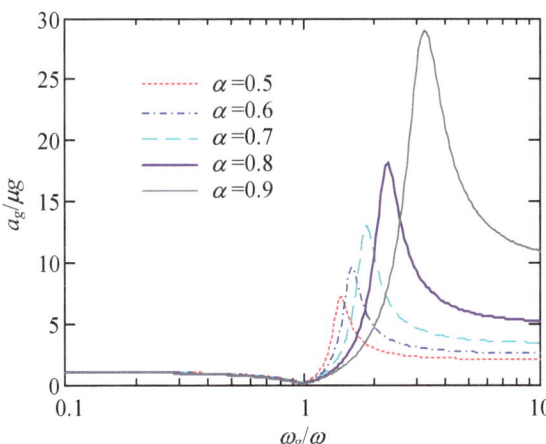

2.3.2 Boundaries Between the Stick-Sliding and Sliding-Sliding Cases

As derived in Sect. 2.2, Eq. (2.10) is the governing equation of the relative displacement during the sliding phase. Equation (2.10) is the differential equation of a SDOF fixed base system with the natural frequency of ω_1 and damping ratio of ξ_1 subjected to a step force. The static displacement corresponding to this step force, u_{st}, is given in Eq. (2.12), which has the same sign as the sliding direction. As shown in Fig. 2.4, for the sliding-sliding case, the sliding base system slides in one direction for half period of the ground motion, $0.5T_g$, and then slides in the opposite direction for another $0.5T_g$, and so on. Therefore, the equivalent step force also changes direction in $0.5T_g$ when the sliding direction changes, as shown in Fig. 2.6. Thus, the relative displacements of the two opposite half sliding cycles have the same absolute values but opposite signs.

For a certain half sliding cycle, the solution of Eq. (2.10) (i.e. Eq. 2.13) can be rewritten as

$$\begin{pmatrix} u_r(t) \\ \dot{u}_r(t) \end{pmatrix} = \mathbf{A}(\tau) \begin{pmatrix} u_{r0} \\ \dot{u}_{r0} \end{pmatrix} + \mathbf{b}(\tau) \tag{2.27}$$

where u_{r0} and $\dot{u}_{r0}$ are the relative displacement and velocity at the moment when this half sliding cycle starts, and this moment is denoted as t_i; $\tau = t - t_i$ is the local time during this half sliding cycle, so $0 \leq \tau \leq \pi/\omega_g$. When $\tau = \pi/\omega_g$, the opposite half sliding cycle starts, the relative displacement and velocity at this moment are equal to $-u_{r0}$ and $-\dot{u}_{r0}$, respectively. Therefore,

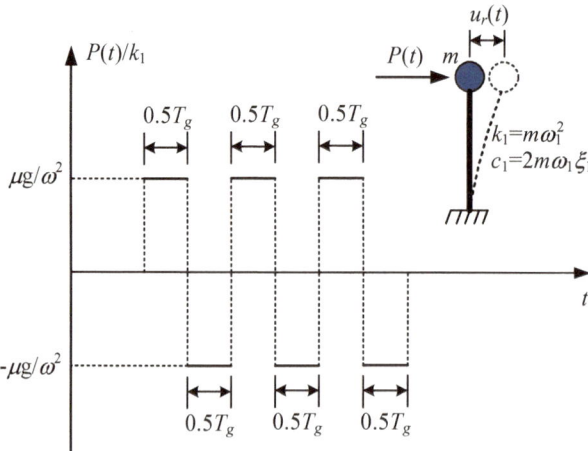

Fig. 2.6 Dynamic model for the relative displacement

$$-\begin{pmatrix} u_{r0} \\ \dot{u}_{r0} \end{pmatrix} = \mathbf{A}(\pi/\omega_g)\begin{pmatrix} u_{r0} \\ \dot{u}_{r0} \end{pmatrix} + \mathbf{b}(\pi/\omega_g) \tag{2.28}$$

The solutions of Eq. (2.28) are

$$u_{r0} = u_{st}\frac{-\sinh\theta_1 + \left(\xi_1/\sqrt{1-\xi_1^2}\right)\sin\theta_2}{\cosh\theta_1 + \cos\theta_2}$$

$$= \frac{-\delta\mu g}{\omega^2}\frac{-\sinh\theta_1 + \left(\xi_1/\sqrt{1-\xi_1^2}\right)\sin\theta_2}{\cosh\theta_1 + \cos\theta_2} \tag{2.29}$$

$$\dot{u}_{r0} = -u_{st}\omega_1\frac{\sin\theta_2/\sqrt{1-\xi_1^2}}{\cosh\theta_1 + \cos\theta_2} = \frac{\delta\mu g\omega_1}{\omega^2}\frac{\sin\theta_2/\sqrt{1-\xi_1^2}}{\cosh\theta_1 + \cos\theta_2} \tag{2.30}$$

where

$$\theta_1 = \xi_1\frac{\omega_1}{\omega_g}\pi, \ \theta_2 = \sqrt{1-\xi_1^2}\frac{\omega_1}{\omega_g}\pi = \frac{\omega_{1d}}{\omega_g}\pi \tag{2.31}$$

and δ represents the direction of the friction force, i.e., the opposite direction of sliding.

By substituting Eqs. (2.29) and (2.30) into Eq. (2.27), the relative displacement and velocity at any time of this half sliding cycle are obtained:

$$u_r(t_i + \tau) = \frac{-\delta\mu g}{\omega^2}$$

$$\left(\frac{1 - e^{-\xi_1\omega_1\tau}}{\frac{\cos(\theta_2 - \omega_{1d}\tau)\sqrt{1-\xi_1^2} - \xi_1\sin(\theta_2 - \omega_{1d}\tau) + \xi_1\sin(\omega_{1d}\tau)e^{\theta_1} + \sqrt{1-\xi_1^2}\cos(\omega_{1d}\tau)e^{\theta_1}}{\sqrt{1-\xi_1^2}(\cosh\theta_1 + \cos\theta_2)}}\right) \tag{2.32}$$

$$\dot{u}_r(t_i + \tau) = \frac{-\delta\mu g\omega_1}{\omega^2}e^{-\xi_1\omega_1\tau}\frac{\sin(\omega_{1d}\tau)\left(e^{\theta_1} + \cos\theta_2\right) - \cos(\omega_{1d}\tau)\sin\theta_2}{\sqrt{1-\xi_1^2}(\cosh\theta_1 + \cos\theta_2)} \tag{2.33}$$

When $\tau = \pi/\omega_g$, the opposite half sliding cycle initiates. From Eq. (2.17), we have

$$\ddot{u}_s(t_i + \pi/\omega_g) = \delta\mu g\pi/\omega_g - \alpha(-2\dot{u}_{r0}) - 2a_g/\omega_g\cos(\omega_g t_i) = 0 \tag{2.34}$$

Therefore,

$$\cos(\omega_g t_i) = \frac{\delta \pi \mu g}{2a_g} + \frac{\alpha \omega_g \dot{u}_{r0}}{a_g} = A_1 \frac{\delta \mu g}{a_g} \tag{2.35}$$

where

$$A_1 = \frac{\alpha \omega_g \omega_1}{\omega^2 \sqrt{1 - \xi_1^2}} \frac{\sin \theta_2}{\cosh \theta_1 + \cos \theta_2} + \frac{\pi}{2} \tag{2.36}$$

Since a certain half sliding cycle initiates at $t = t_i$,

$$|\alpha \ddot{u}_r + \ddot{u}_g|_{t=t_i} = \delta \big(\alpha \big(-2\xi \omega \dot{u}_{r0} - \omega^2 u_{r0} - a_g \sin(\omega_g t_i)\big) + a_g \sin(\omega_g t_i)\big)$$
$$= -A_2 \mu g + \delta(1 - \alpha) a_g \sin(\omega_g t_i) \geq \mu g \tag{2.37}$$

where

$$A_2 = \frac{\alpha \Big(\xi_1 \sin \theta_2 / \sqrt{1 - \xi_1^2} + \sinh \theta_1\Big)}{\cosh \theta_1 + \cos \theta_2} \tag{2.38}$$

From Eqs. (2.35) and (2.37), we have

$$(1 - \alpha)^2 \big(a_g^2 - A_1^2 (\mu g)^2\big) \geq (1 + A_2)^2 (\mu g)^2 \tag{2.39}$$

Thus, the boundaries between the stick-sliding and sliding-sliding cases are

$$\frac{a_g}{\mu g} = \sqrt{(1 + A_2)^2 / (1 - \alpha)^2 + A_1^2} \tag{2.40}$$

The variations of the boundaries between the stick-sliding and sliding-sliding cases for different ω_g / ω and α are shown in Fig. 2.7.

When $\omega_g / \omega_1 \to 0$, θ_1 and $\theta_2 \to +\infty$, $A_1 \to \pi/2$, and $A_2 \to \alpha$; thus, Eq. (2.40) becomes

$$\frac{a_g}{\mu g} = \sqrt{(1 + \alpha)^2 / (1 - \alpha)^2 + (\pi/2)^2} \tag{2.41}$$

When $\omega_g / \omega_1 \to +\infty$, θ_1 and $\theta_2 \to 0$, $A_1 \to \pi/(2(1 - \alpha))$, and $A_2 \to 0$; thus, Eq. (2.40) becomes

$$\frac{a_g}{\mu g} = \frac{\sqrt{1 + (\pi/2)^2}}{1 - \alpha} \tag{2.42}$$

Fig. 2.7 Boundaries between the stick-sliding and sliding-sliding cases ($\xi = 5\%$)

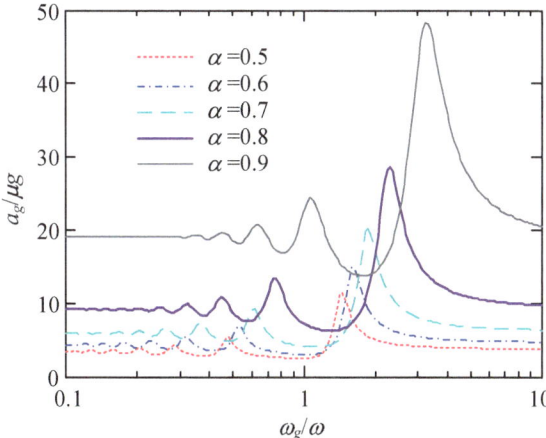

As can be seen in Fig. 2.7, there are also some local peak values when $\theta_2 = (2n - 1)\pi$. These points are corresponding to the resonance.

2.4   Parametric Study for the Maximum Responses

2.4.1   Maximum Pseudo Acceleration of the Top Mass

The maximum shear force applied to the superstructure can be represented by the maximum pseudo acceleration of the top mass (Chopra, 2001), A, as follows:

$$A = \omega^2 \times (\max|u_r(t)|) = \frac{k}{m}(\max|u_r(t)|) \tag{2.43}$$

The function of a_g, μg, ω_g/ω, α, and ξ is related to A in the 2DOF sliding base system depicted in Fig. 2.1. Figure 2.8 illustrates how the normalized maximum pseudo acceleration, $A/\mu g$, relates to the frequency ratio, ω_g/ω, for varying values of α and $a_g/\mu g$. A sliding base structure, unlike fixed base structures, has multiple resonant frequencies when sliding occurs. $a_g/\mu g$ and α can also affect these resonant frequencies. As shown in Fig. 2.8a, b, as $a_g/\mu g$ increases, several resonant frequency ratios appear in the region of $\omega_g/\omega < 1$, and these ratios shift towards higher values gradually and eventually reach upper limits, which correspond to the sliding-sliding cases. As shown in Fig. 2.8c, d, resonances are more prone to happen with smaller mass ratios, α.

As shown in Fig. 2.8c, d, the maximum pseudo acceleration typically reduces with an increase in the mass ratio, α. The reason for this is that the damping ratio corresponding to the sliding phase, ξ_1, is equal to $\xi/\sqrt{1 - \alpha}$ as given in Eq. (2.11), which increases as α increases. However, in the resonant frequency ranges of a

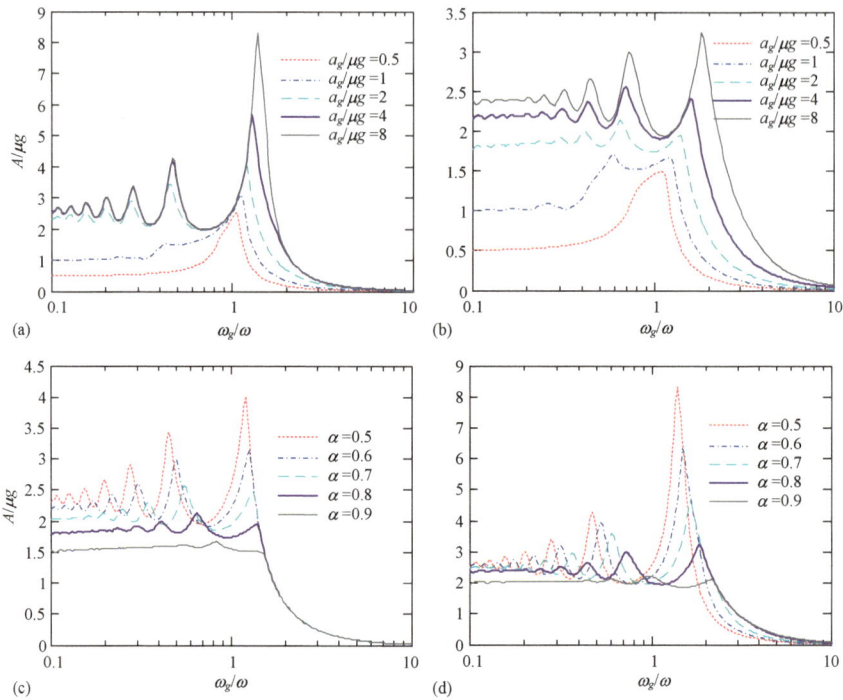

Fig. 2.8 Relationship between the normalized maximum pseudo acceleration and ω_g/ω ($\xi = 5\%$): **a** $\alpha = 0.5$; **b** $\alpha = 0.8$; **c** $a_g/\mu g = 2$; and **d** $a_g/\mu g = 8$

specific sliding base structure, the maximum pseudo acceleration may be greater compared to other structures with lower mass ratios.

Figure 2.9 shows the relationship between $A/\mu g$ and $a_g/\mu g$ for different values of α and ω_g/ω, where the circle and triangle represent the boundary between the stick-stick and stick-sliding cases, and the boundary between the stick-sliding and sliding-sliding cases, respectively. The $A/\mu g$ versus $a_g/\mu g$ curves are segregated into three areas, each corresponding to cases of stick-stick, stick-sliding, and sliding-sliding. In the stick-stick region, responses remain independent of mass ratio, α, with the $A/\mu g$ versus $a_g/\mu g$ curve taking the form of a straight line. The slope of this straight line is equal to the displacement response factor, R_d, for a fixed base structure:

$$R_d = \frac{1}{\sqrt{\left[1 - (\omega_g/\omega)^2\right]^2 + \left[2\xi(\omega_g/\omega)\right]^2}} \tag{2.44}$$

In the stick-sliding region, due to additional resonances when $\omega_g/\omega < 1$, the maximum pseudo acceleration of a sliding base structure may exceed that of the related fixed base structure. This phenomenon can be observed in Fig. 2.9a for $\omega_g/\omega = 0.1$ and 0.5. In the sliding-sliding region, the upper limit of the maximum pseudo

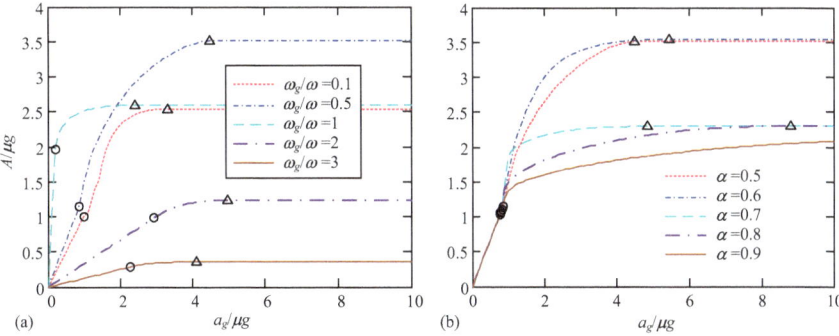

(a)   (b)

o Boundary between the stick-stick and stick-sliding cases
△ Boundary between the stick-sliding and sliding-sliding cases

Fig. 2.9 Relationship between the normalized maximum pseudo acceleration and $a_g/\mu g$ ($\xi = 5\%$): **a** $\alpha = 0.5$; **b** $\omega_g/\omega = 0.5$

acceleration is reached, and does not change even when there are larger ground accelerations. This upper limit reflects the effectiveness of sliding base structures for isolating extremely large earthquakes. As shown in Fig. 2.9b, in normal circumstances, as the mass ratio increases, the likelihood of the sliding-sliding case occurring becomes more difficult. This trend can be more clearly observed in Fig. 2.7.

2.4.2 Amplitude of the Sliding Displacement

As shown in Fig. 2.4b, the maximum sliding displacement results from the accumulation of transient responses that occur before reaching the steady periodic state. Hence, the maximum sliding displacement is considerably influenced by the initial ground acceleration, e.g., there is a significant disparity in the maximum sliding displacement between the sine and cosine ground accelerations. To represent the extent of sliding when exposed to harmonic ground motions, the amplitude of the sliding displacement, $u_{s,ap}$, is a more appropriate response quantity compared to the maximum sliding displacement. It is defined as the difference between the maximum and minimum sliding displacements during the steady state; therefore, the value of it is exclusively linked to the responses of the steady state.

Figure 2.10 depicts the correlation between the frequency ratio, ω_g/ω, and the normalized sliding displacement amplitude, $u_{s,ap}/(a_g/\omega_g^2)$, for varying α and $a_g/\mu g$ values. As shown in Fig. 2.10a, b, when ω_g/ω A is less than 1, the normalized sliding displacement amplitude is not greatly affected by the mass, α, ratio or frequency ratio, ω_g/ω. There is a noticeable reduction of the sliding displacement amplitude when ω_g/ω increases, after the frequency ratio, ω_g/ω, surpasses 1. This result is in agreement with Fig. 2.5, where the critical $a_g/\mu g$ for the occurrence of the stick-sliding case increases as ω_g/ω increases once ω_g/ω surpasses 1. As shown in Fig. 2.10b, if

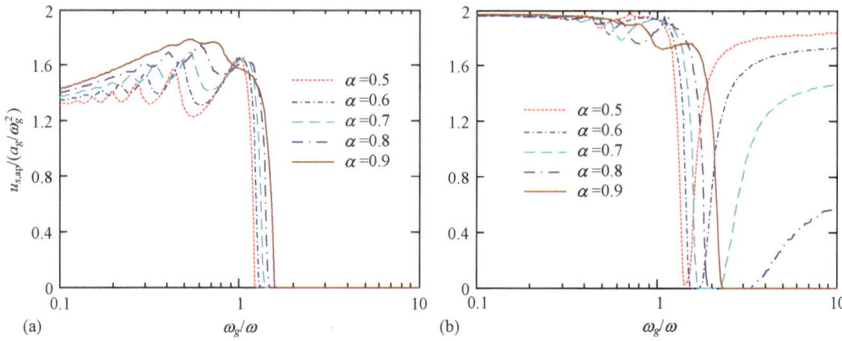

Fig. 2.10 Relationship between the normalized sliding displacement amplitude and ω_g/ω ($\xi = 5\%$): **a** $a_g/\mu g = 2$; and **b** $a_g/\mu g = 8$

the normalized ground acceleration amplitude, $a_g/\mu g$, is sufficiently large, as ω_g/ω surpasses a certain value, the sliding displacement amplitude will rise alongside an increase in ω_g/ω. Figure 2.5 is also in accordance with this, where the critical value of $a_g/\mu g$ leading to stick-sliding decreases with increasing value of ω_g/ω once the peak point of critical value $a_g/\mu g$ is reached. Figure 2.10b also indicates that there is a tendency generally for the sliding displacement amplitude to decrease as the mass ratio increases in the area of large frequency ratios.

Figure 2.11 depicts the correlation between the normalized sliding displacement amplitude, $u_{s,ap}/\left(a_g/\omega_g^2\right)$, and the normalized ground acceleration amplitude, $a_g/\mu g$, for different ω_g/ω and α. As shown in Fig. 2.11, after sliding occurs, the normalized sliding displacement amplitude increases as $a_g/\mu g$ increases, but with a gradually decreasing speed increase. The normalized sliding displacement amplitude has an upper bound value that is 2. The reason for this is that when $a_g/\mu g$ is very large, the absolute acceleration of the sliding base (which is limited by the friction coefficient) is negligible compared with the ground acceleration, so the sliding base can be considered motionless from the perspective of ground, and $u_{s,ap}$ is equal to the vibration amplitude of the ground displacement, $2a_g/\omega_g^2$.

2.5 Theoretical Solutions for the Responses of the Sliding-Sliding Case

For the sliding-sliding case, the maximum pseudo acceleration reaches the upper limit, and is no longer dependent on the amplitude of the ground acceleration. This upper limit response has important physical meaning, and reflects the effectiveness of the sliding base system for reducing the superstructure response.

In this section, the theoretical solutions for the maximum pseudo acceleration and sliding displacement amplitude for the sliding-sliding case are derived. The derived results are further used to explain the mechanism of the sliding base system.

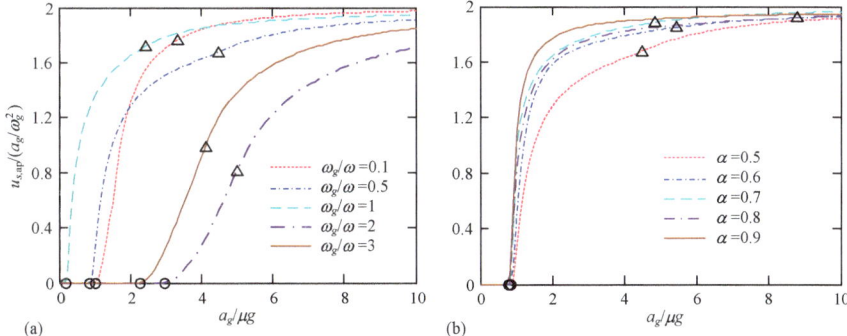

Fig. 2.11 Relationship between the normalized sliding displacement amplitude and $a_g/\mu g$ ($\xi = 5\%$): **a** $\alpha = 0.5$; **b** $\omega_g/\omega = 0.5$

For the stick-sliding case, theoretical solutions cannot be obtained because of the complicated transformation between stick and sliding phases.

2.5.1 Solutions for the Maximum Pseudo Acceleration

The maximum relative displacement will be obtained when the relative velocity is equal to 0, i.e. $\dot{u}_r = 0$; thus, the corresponding local time, τ_j, can be obtained by taking the right-hand side of Eq. (2.33) equal to zero:

$$\tan(\omega_{1d}\tau_j) = \frac{\sin\theta_2}{e^{\theta_1} + \cos\theta_2} \tag{2.45}$$

From Eq. (2.45), several solutions for τ_j may be obtained, and the number of solutions is dependent on the value of θ_2. By substituting Eq. (2.45) into Eq. (2.32), the relative displacement at the moment of τ_j can be obtained:

$$u_r\left(t_i + \tau_j\right) = \frac{-\delta\mu g}{\omega^2}\left(1 - e^{-\frac{\xi_1}{\sqrt{1-\xi_1^2}}\omega_{1d}\tau_j}\frac{\cos\left(\theta_2 - \omega_{1d}\tau_j\right) + \cos\left(\omega_{1d}\tau_j\right)e^{\theta_1}}{\cosh\theta_1 + \cos\theta_2}\right) \tag{2.46}$$

Using Eq. (2.46), the normalized maximum pseudo acceleration can be written as

$$A/\mu g = \max \left| \frac{u_r(t_i + \tau_j)\omega^2}{\mu g} \right|$$

$$= \max \left| 1 - e^{-\frac{\xi_1}{\sqrt{1-\xi_1^2}} \omega_{1d}\tau_j} \frac{\cos(\theta_2 - \omega_{1d}\tau_j) + \cos(\omega_{1d}\tau_j)e^{\theta_1}}{\cosh \theta_1 + \cos \theta_2} \right| \quad (2.47)$$

Equation (2.47) is not an explicit equation yet, since there may be more than one solution for τ_j and the corresponding pseudo acceleration. The maximum pseudo acceleration is the maximum of these candidates. To obtain the explicit form of Eq. (2.47), different cases of θ_2 need to be considered. The following presents the complete derivations for the explicit expressions of the normalized maximum pseudo acceleration, $A/\mu g$.

Equation (2.45) leads to

$$\sin^2(\omega_{1d}\tau_j) = \frac{\tan^2(\omega_{1d}\tau_j)}{\tan^2(\omega_{1d}\tau_j) + 1} = \frac{\sin^2 \theta_2}{2e^{\theta_1}(\cosh \theta_1 + \cos \theta_2)} \quad (2.48)$$

If $\theta_2 \neq n\pi$ ($n = 1, 2, \ldots$), $\sin \theta_2 \neq 0$, so $\sin(\omega_{1d}\tau_j) \neq 0$ from Eq. (2.48). Therefore, Eq. (2.47) can be rewritten as Eq. (2.49) by using Eq. (2.45).

$$A/\mu g = \max \left| 1 - e^{-\frac{\xi_1}{\sqrt{1-\xi_1^2}} \omega_{1d}\tau_j} \frac{\sin \theta_2}{\sin(\omega_{1d}\tau_j)(\cosh \theta_1 + \cos \theta_2)} \right| \quad (2.49)$$

Case 1. If $0 < \theta_2 < \pi$, $\sin \theta_2 > 0$. From Eq. (2.45), we have

$$\tan(\omega_{1d}\tau_j) = \frac{\sin \theta_2}{e^{\theta_1} + \cos \theta_2} < \frac{\sin \theta_2}{1 + \cos \theta_2} = \tan(\theta_2/2) \quad (2.50)$$

Since $0 \leq \tau_j \leq \pi/\omega_g$, $0 \leq \omega_{1d}\tau_j \leq \theta_2$, so there is only one solution for τ_j, and

$$\omega_{1d}\tau_j = \arctan\left(\frac{\sin \theta_2}{e^{\theta_1} + \cos \theta_2}\right) < \theta_2/2 < \pi/2 \quad (2.51)$$

From Eq. (2.51), we have $\sin(\omega_{1d}\tau_j) > 0$, so from Eq. (2.48), we have

$$\frac{\sin \theta_2}{\sin(\omega_{1d}\tau_j)} = \sqrt{2e^{\theta_1}(\cosh \theta_1 + \cos \theta_2)} \quad (2.52)$$

Substituting Eq. (2.52) into Eq. (2.49) gives

$$A/\mu g = \left| 1 - e^{-\frac{\xi_1}{\sqrt{1-\xi_1^2}}\omega_{1d}\tau_j} \sqrt{\frac{2e^{\theta_1}}{\cosh\theta_1 + \cos\theta_2}} \right| \quad (2.53)$$

Case 2. If $\pi < \theta_2 < 2\pi$, $\sin\theta_2 < 0$, From Eq. (2.45), we have

$$\tan(\theta_2/2) = \frac{\sin\theta_2}{1 + \cos\theta_2} < \tan(\omega_{1d}\tau_j) = \frac{\sin\theta_2}{e^{\theta_1} + \cos\theta_2} < 0 \quad (2.54)$$

so the smallest solution for τ_j is

$$\pi/2 < \theta_2/2 < \omega_{1d}\tau_j = \arctan\left(\frac{\sin\theta_2}{e^{\theta_1} + \cos\theta_2}\right) + \pi < \pi \quad (2.55)$$

Since

$$\arctan\left(\frac{\sin\theta_2}{e^{\theta_1} + \cos\theta_2}\right) + 2\pi > \theta_2/2 + \pi > \theta_2 \quad (2.56)$$

there is no other solutions for τ_j because $0 \leq \omega_{1d}\tau_j \leq \theta_2$. From Eq. (2.55), we have $\sin(\omega_{1d}\tau_j) > 0$, so from Eq. (2.48), we have

$$\frac{\sin\theta_2}{\sin(\omega_{1d}\tau_j)} = -\sqrt{2e^{\theta_1}(\cosh\theta_1 + \cos\theta_2)} \quad (2.57)$$

Substituting Eq. (2.57) into Eq. (2.49) gives

$$A/\mu g = 1 + e^{-\frac{\xi_1}{\sqrt{1-\xi_1^2}}\omega_{1d}\tau_j} \sqrt{\frac{2e^{\theta_1}}{\cosh\theta_1 + \cos\theta_2}} \quad (2.58)$$

Case 3. If $\theta_2 > 2\pi$ and $\theta_2 \neq n\pi$ $(n = 1, 2, \ldots)$, there are at least two solutions for τ_j. If $\sin\theta_2 > 0$, i.e., $2n\pi < \theta_2 < (2n+1)\pi$, then from Eq. (2.45), we have $\tan(\omega_{1d}\tau_j) > 0$. Therefore, the first solution for τ_j is

$$0 < \omega_{1d}\tau_j = \arctan\left(\frac{\sin\theta_2}{e^{\theta_1} + \cos\theta_2}\right) < \pi/2 \quad (2.59)$$

and the corresponding pseudo acceleration is

$$\left| \frac{u_r(t_i + \tau_j)\omega^2}{\mu g} \right| = \left| 1 - e^{-\frac{\xi_1}{\sqrt{1-\xi_1^2}}\phi} \sqrt{\frac{2e^{\theta_1}}{\cosh\theta_1 + \cos\theta_2}} \right| \quad (2.60)$$

where

$$\phi = \arctan\left(\frac{\sin\theta_2}{e^{\theta_1} + \cos\theta_2}\right) \tag{2.61}$$

The second solution for τ_j is

$$\pi < \omega_{1d}\tau_j = \arctan\left(\frac{\sin\theta_2}{e^{\theta_1} + \cos\theta_2}\right) + \pi < 3\pi/2 \tag{2.62}$$

and the corresponding pseudo acceleration is

$$\left|\frac{u_r(t_i + \tau_j)\omega^2}{\mu g}\right| = 1 + e^{-\frac{\xi_1}{\sqrt{1-\xi_1^2}}(\phi+\pi)}\sqrt{\frac{2e^{\theta_1}}{\cosh\theta_1 + \cos\theta_2}} \tag{2.63}$$

The pseudo accelerations corresponding to other solutions of τ_j are all smaller than the larger one of Eqs. (2.60) and (2.63), since the oscillation of the relative displacement gradually decays because of the damping. If Eq. (2.60) is equal to

$$\left|\frac{u_r(t_i + \tau_j)\omega^2}{\mu g}\right| = 1 - e^{-\frac{\xi_1}{\sqrt{1-\xi_1^2}}\phi}\sqrt{\frac{2e^{\theta_1}}{\cosh\theta_1 + \cos\theta_2}} \tag{2.64}$$

it is clear that Eq. (2.63) is larger than Eq. (2.60). If Eq. (2.60) is equal to

$$\left|\frac{u_r(t_i + \tau_j)\omega^2}{\mu g}\right| = e^{-\frac{\xi_1}{\sqrt{1-\xi_1^2}}\phi}\sqrt{\frac{2e^{\theta_1}}{\cosh\theta_1 + \cos\theta_2}} - 1 \tag{2.65}$$

Equation (2.60) minus Eq. (2.63) is

$$
\begin{aligned}
& e^{-\frac{\xi_1}{\sqrt{1-\xi_1^2}}\phi}\sqrt{\frac{2e^{\theta_1}}{\cosh\theta_1 + \cos\theta_2}}\left(1 - e^{-\frac{\xi_1}{\sqrt{1-\xi_1^2}}\pi}\right) - 2 \\
& < \sqrt{\frac{2e^{\theta_1}}{\cosh\theta_1 + \cos\theta_2}}\left(1 - e^{-\frac{\theta_1}{\theta_2}\pi}\right) - 2 \\
& < \sqrt{\frac{2e^{\theta_1}}{\cosh\theta_1 - 1}}\left(1 - e^{-\theta_1}\right) - 2 \\
& = \frac{e^{\theta_1} - 1}{\cosh\theta_1 - 1}\left(1 - e^{-\theta_1}\right) - 2 = 0
\end{aligned}
\tag{2.66}
$$

Therefore, Eq. (2.63) is always larger than Eq. (2.60), and the maximum pseudo acceleration is obtained at

$$\pi < \omega_{1d}\tau_j = \arctan\left(\frac{\sin\theta_2}{e^{\theta_1} + \cos\theta_2}\right) + \pi < 3\pi/2 \tag{2.67}$$

and the maximum pseudo acceleration is

$$A/\mu g = 1 + e^{-\frac{\xi_1}{\sqrt{1-\xi_1^2}}\omega_{1d}\tau_j} \sqrt{\frac{2e^{\theta_1}}{\cosh\theta_1 + \cos\theta_2}} \tag{2.68}$$

Case 4. If $\sin\theta_2 < 0$, i.e., $(2n+1)\pi < \theta_2 < (2n+2)\pi$, then from Eq. (2.45), we have $\tan(\omega_{1d}\tau_j) < 0$. Therefore, the maximum pseudo acceleration will be obtained at the first solution of τ_j, that is

$$\pi/2 < \omega_{1d}\tau_j = \arctan\left(\frac{\sin\theta_2}{e^{\theta_1} + \cos\theta_2}\right) + \pi < \pi \tag{2.69}$$

and the maximum pseudo acceleration is

$$A/\mu g = 1 + e^{-\frac{\xi_1}{\sqrt{1-\xi_1^2}}\omega_{1d}\tau_j} \sqrt{\frac{2e^{\theta_1}}{\cosh\theta_1 + \cos\theta_2}} \tag{2.70}$$

Case 5. If $\theta_2 = n\pi$ $(n = 1, 2, \ldots)$, $\sin\theta_2 = 0$, and from Eq. (2.45), we have $\tan(\omega_{1d}\tau_j) = 0$, so the first solution for τ_j is

$$\omega_{1d}\tau_j = 0 \tag{2.71}$$

Substituting Eq. (2.71) into Eq. (2.46) gives the corresponding pseudo acceleration:

$$\left|\frac{u_r(t_i + \tau_j)\omega^2}{\mu g}\right| = \frac{(-1)^n + e^{\theta_1}}{\cosh\theta_1 + (-1)^n} - 1 \tag{2.72}$$

The second solution for τ_j is

$$\omega_{1d}\tau_j = \pi \tag{2.73}$$

and Substituting Eq. (2.73) into Eq. (2.46) gives the corresponding pseudo acceleration:

$$\left|\frac{u_r(t_i + \tau_j)\omega^2}{\mu g}\right| = 1 + e^{-\frac{\xi_1}{\sqrt{1-\xi_1^2}}\pi} \frac{(-1)^n + e^{\theta_1}}{\cosh\theta_1 + (-1)^n}$$

$$= 1 + e^{-\theta_1/n} \frac{(-1)^n + e^{\theta_1}}{\cosh\theta_1 + (-1)^n} \tag{2.74}$$

Equation (2.72) minus Eq. (2.74) is

$$\frac{(-1)^n + e^{\theta_1}}{\cosh \theta_1 + (-1)^n}\left(1 - e^{-\theta_1/n}\right) - 2 \le \frac{(-1)^n + e^{\theta_1}}{\cosh \theta_1 + (-1)^n}\left(1 - e^{-\theta_1}\right) - 2$$
$$= \begin{cases} \frac{-2(1+e^{\theta_1})}{\cosh \theta_1 + 1} < 0 \ (n \text{ is even}) \\ 0 \qquad\qquad\qquad (n \text{ is odd}) \end{cases} \tag{2.75}$$

Therefore, Eq. (2.72) is always smaller than Eq. (2.74). For $\theta_2 = n\pi \ (n = 1, 2, \ldots)$,

$$\frac{(-1)^n + e^{\theta_1}}{\cosh \theta_1 + (-1)^n} = \sqrt{\frac{2e^{\theta_1}}{\cosh \theta_1 + (-1)^n}} = \sqrt{\frac{2e^{\theta_1}}{\cosh \theta_1 + \cos \theta_2}} \tag{2.76}$$

so the maximum pseudo acceleration can be written as

$$A/\mu g = 1 + e^{-\frac{\xi_1}{\sqrt{1-\xi_1^2}}\pi}\sqrt{\frac{2e^{\theta_1}}{\cosh \theta_1 + \cos \theta_2}} \tag{2.77}$$

The results of Cases 2–5 can be combined in one equation, that is, for $\theta_2 \ge \pi$, the normalized maximum pseudo acceleration is

$$A/\mu g = 1 + e^{-\frac{\xi_1}{\sqrt{1-\xi_1^2}}\omega_{1d}\tau_j}\sqrt{\frac{2e^{\theta_1}}{\cosh \theta_1 + \cos \theta_2}} \tag{2.78}$$

where

$$\pi/2 < \omega_{1d}\tau_j = \arctan\left(\frac{\sin \theta_2}{e^{\theta_1} + \cos \theta_2}\right) + \pi < 3\pi/2 \tag{2.79}$$

In summary, the explicit solutions for the maximum pseudo acceleration are as follows:

For $0 < \theta_2 < \pi$, the normalized maximum pseudo acceleration is

$$A/\mu g = \left| 1 - e^{-\frac{\xi_1}{\sqrt{1-\xi_1^2}}\omega_{1d}\tau_0}\sqrt{\frac{2e^{\theta_1}}{\cosh \theta_1 + \cos \theta_2}} \right| \tag{2.80}$$

where

$$\omega_{1d}\tau_0 = \arctan\left(\frac{\sin \theta_2}{e^{\theta_1} + \cos \theta_2}\right) < \pi/2 \tag{2.81}$$

For $\theta_2 \geq \pi$, the normalized maximum pseudo acceleration is

$$A/\mu g = 1 + e^{-\frac{\xi_1}{\sqrt{1-\xi_1^2}}\omega_{1d}\tau_0} \sqrt{\frac{2e^{\theta_1}}{\cosh\theta_1 + \cos\theta_2}} \tag{2.82}$$

where

$$\pi/2 < \omega_{1d}\tau_0 = \arctan\left(\frac{\sin\theta_2}{e^{\theta_1} + \cos\theta_2}\right) + \pi < 3\pi/2 \tag{2.83}$$

In the above equations, τ_0 is the local time corresponding to the maximum pseudo acceleration. θ_1 and θ_2 in Eqs. (2.80)–(2.83) are only related to ξ_1 and ω_1/ω_g as given in Eq. (2.31); thus, the normalized maximum pseudo acceleration, $A/\mu g$, is only dependent on ξ_1 and ω_1/ω_g for the sliding-sliding case. ξ_1 and ω_1 are the natural frequency and damping ratio for the sliding phase, which are related to the mass ratio, α, as shown in Eq. (2.11).

2.5.2 Interpretation of the Solutions for the Maximum Pseudo Acceleration

Figure 2.12 plots the relationship between $A/\mu g$ and ω_g/ω using the derived theoretical solutions. As shown in Fig. 2.12a, the results from the theoretical solutions and numerical methods are the same, which verifies the accuracy of the theoretical solutions. As shown in Fig. 2.12b, as the mass ratio, α, increases, the frequency ratios, ω_g/ω, of resonance shift towards larger values and the general responses decrease. This is because ξ_1 and ω_1 increase as α increases for given ξ and ω.

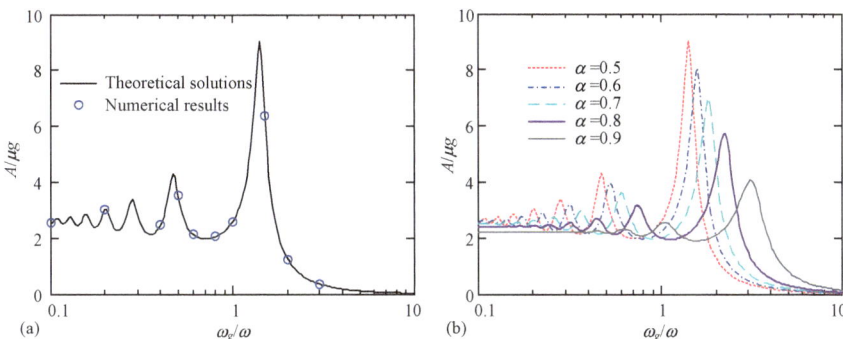

Fig. 2.12 Relationship between $A/\mu g$ and ω_g/ω ($\xi = 5\%$): **a** verification of the theoretical solutions for $\alpha = 0.5$; **b** results for different values of α

In the following paragraphs, we will examine various frequency ratio regions, and elucidate the outcomes illustrated in Fig. 2.12b. When $\omega_g/\omega_1 \to 0$, θ_1 and $\theta_2 \to +\infty$, so from Eq. (2.83), we have $\omega_{1d}\tau_0 \to \pi$, and Eq. (2.82) becomes

$$A/\mu g = 1 + e^{-\frac{\xi_1}{\sqrt{1-\xi_1^2}}\omega_{1d}\tau_0} \sqrt{\frac{2e^{\theta_1}}{\cosh\theta_1 + \cos\theta_2}} \to 1 + 2e^{-\frac{\xi_1}{\sqrt{1-\xi_1^2}}\pi} \tag{2.84}$$

If the damping ratio, $\xi = 0$, the right-hand side of Eq. (2.84) is equal to 3, which implies that the maximum pseudo acceleration is equal to 3 times μg. For actual structures with damping, the normalized maximum pseudo acceleration, $A/\mu g$, is always between 1 and 3, as demonstrated by Eq. (2.84). This is because when $\omega_g/\omega_1 \to 0$, the ground motion period is much larger than the natural period of the sliding system, so there is sufficient time for the oscillation of the relative displacement to decrease and stabilize to the static displacement, $u_{st} = -\delta\mu g/\omega^2$, prior to the onset of the opposite sliding. Once sliding changes direction, the relative displacement initiates oscillation around the new static equilibrium displacement, $-u_{st}$. If the damping is sufficiently small, the relative displacement can approach $-3u_{st}$ during the first oscillation cycle; if the damping is very large, the oscillation of the relative displacement diminishes rapidly and is only capable of attaining the new static equilibrium displacement, $-u_{st}$. Figure 2.13 shows the steady state response of the normalized pseudo acceleration for $\alpha = 0.8$, $\xi = 5\%$, $T = 1$ s and $T_g = 10$ s ($\omega_g/\omega_1 = 0.045$). Equation (2.82) can also be written as

$$A/\mu g = 1 + 2e^{-\frac{\xi_1}{\sqrt{1-\xi_1^2}}\omega_{1d}\tau_0} \sqrt{\frac{1}{\left(e^{-\theta_1}-1\right)^2 + 2e^{-\theta_1}(\cos\theta_2+1)}} \tag{2.85}$$

If $\xi = 0$ [which means $\theta_1 = 0$ from Eq. (2.31)] and $\theta_2 = (2n-1)\pi$ (where n is a positive integer), the normalized maximum pseudo acceleration, $A/\mu g$, becomes infinite from Eq. (2.85). If $\xi \neq 0$, a local peak value will be obtained for $A/\mu g$ when $\theta_2 = (2n-1)\pi$, i.e., $\omega_{1d} = (2n-1)\omega_g$ from Eq. (2.31), in which ω_{1d} is the natural frequency of the damped vibration during the sliding phase:

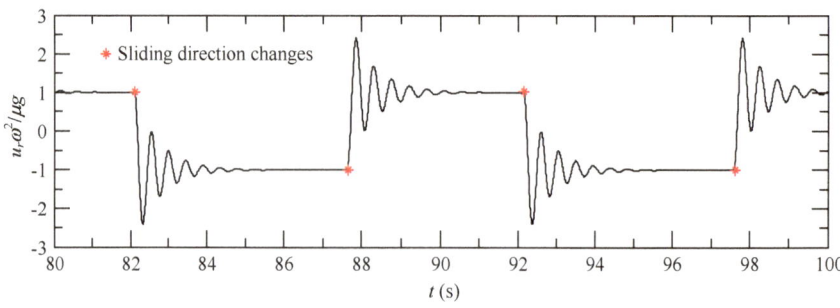

Fig. 2.13 Normalized pseudo acceleration response for $\alpha = 0.8$, $T = 1$ s, $T_g = 10$ s and $\xi = 5\%$

$$A/\mu g = 1 + 2e^{-\frac{\xi_1}{\sqrt{1-\xi_1^2}}\pi}\frac{1}{1-e^{-\theta_1}} \tag{2.86}$$

This result is consistent with the observations from Fig. 2.12b that there are multiple resonant frequencies in the area of $\omega_g/\omega_1 < 1$.

For $\xi = 0$, Fig. 2.14 displays the normalized pseudo acceleration response for $\omega_1 = \omega_g$ and $\omega_1 = 3\omega_g$. The ground acceleration, $\ddot{u}_g$, is taken as $a_g \cos(\omega_g t)$, where a_g is a large value so that sliding can occur during the whole time of investigation. Since $\omega_{1d} = \omega_1 = (2n-1)\omega_g$ for $\xi = 0$, the relative displacement reaches the peak value of a certain half sliding cycle when the direction of sliding changes, which is the furthest location from the static equilibrium displacement of the next half sliding cycle. From Eq. (2.13), the relative displacement at the end of one half sliding cycle for $\xi = 0$ is

$$u_r(t_i + T_g/2) = \cos(\omega_1 T_g/2)u_r(t_i) + u_{st} - u_{st}\cos(\omega_1 T_g/2) = -u_r(t_i) + 2u_{st} \tag{2.87}$$

So

$$\left|u_r(t_i + T_g/2)\right| - |u_r(t_i)| = 2\mu g/\omega^2 \tag{2.88}$$

Thus, during one cycle of the ground motion, T_g, the maximum relative displacement can increase by $4\mu g/\omega^2$, and can continue increasing until the ground acceleration is not large enough anymore to start sliding. For actual structures, the maximum relative displacement will be reached after several cycles of ground motion because of the damping, as shown in Fig. 2.15 for $\xi = 5\%$, $\alpha = 0.8$, $T = 1$ s and $T_g = 0.45$ s $(\omega_g/\omega_{1d} = 1)$. When $\omega_g/\omega_1 \to 0$, $\theta_1 \to +\infty$ from Eq. (2.31), so Eq. (2.86) tends toward Eq. (2.84). This means that when ω_g/ω decreases, the resonance will slowly disappear, as shown in Fig. 2.12b.

When $\omega_g/\omega_1 \to +\infty$, θ_1 and $\theta_2 \to 0$ from Eq. (2.31), we have $\omega_{1d}\tau_0 \to \theta_2/2$ from Eq. (2.81), so Eq. (2.80) becomes

$$A/\mu g = \left|1 - e^{-\frac{\xi_1}{\sqrt{1-\xi_1^2}}\omega_{1d}\tau_0}\sqrt{\frac{2e^{\theta_1}}{\cosh\theta_1 + \cos\theta_2}}\right| \to \left|\frac{1-2\xi_1^2}{8}\left(\frac{\omega_1}{\omega_g}\pi\right)^2\right| \to 0 \tag{2.89}$$

This is because when $\omega_g/\omega_1 \to +\infty$, the frequency of ground acceleration is considerably larger than the natural frequency of the sliding system, so the oscillation of the relative displacement can be hardly stimulated because of the frequently changed sliding direction.

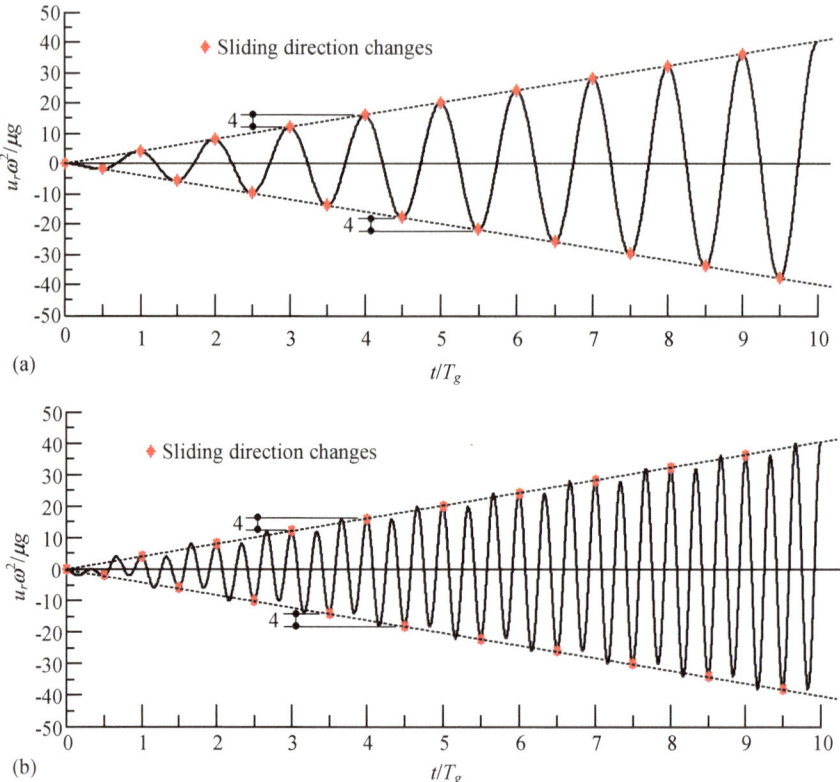

Fig. 2.14 Response of the normalized pseudo acceleration for $\xi = 0$: **a** $\omega_1 = \omega_g$; **b** $\omega_1 = 3\omega_g$

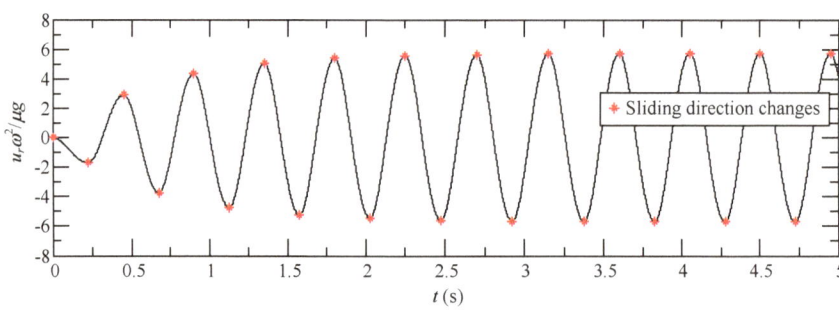

Fig. 2.15 Response of the normalized pseudo acceleration for $\xi = 5\%$, $\alpha = 0.8$, $T = 1$ s and T_g $= 0.45$ s

2.5.3 Solutions for the Sliding Displacement Amplitude

In Sect. 2.4.2, we have obtained

$$
\cos(\omega_g t_i) = \frac{\delta \pi \mu g}{2 a_g} + \frac{\alpha \omega_g \dot{u}_{r0}}{a_g}
$$

$$
= \frac{\delta \mu g}{a_g} \left(\frac{\alpha \omega_g \omega_1}{\omega^2 \sqrt{1 - \xi_1^2}} \frac{\sin \theta_2}{\cosh \theta_1 + \cos \theta_2} + \frac{\pi}{2} \right) \tag{2.90}
$$

By substituting $t = t_i + \pi/\omega_g$ into Eq. (2.18) and using Eq. (2.90), the amplitude of the sliding displacement can be obtained as follows:

$$
u_{s,ap} = \left| u_s(t_i + \pi/\omega_g) - u_s(t_i) \right|
$$

$$
= \left| \frac{1}{2} \delta \mu g (\pi/\omega_g)^2 + \alpha \left(2 u_{r0} + \dot{u}_{r0}(\pi/\omega_g) \right) - \frac{2 a_g}{\omega_g^2} \sin(\omega_g t_i) - \frac{\pi a_g}{\omega_g^2} \cos(\omega_g t_i) \right|
$$

$$
= \left| \frac{-2 a_g}{\omega_g^2} \sin(\omega_g t_i) + 2 \alpha u_{r0} \right|
$$

$$
= \left| \frac{2 a_g}{\omega_g^2} \sin(\omega_g t_i) + 2 \alpha \frac{\delta \mu g}{\omega^2} \frac{-\sinh \theta_1 + \left(\xi_1/\sqrt{1 - \xi_1^2} \right) \sin \theta_2}{\cosh \theta_1 + \cos \theta_2} \right| \tag{2.91}
$$

Therefore, the normalized sliding displacement amplitude is

$$
u_{s,ap} / \left(a_g/\omega_g^2 \right) = \left| 2 \sin(\omega_g t_i) + 2 \alpha \frac{\delta \mu g \omega_g^2}{a_g \omega^2} \frac{-\sinh \theta_1 + \left(\xi_1/\sqrt{1 - \xi_1^2} \right) \sin \theta_2}{\cosh \theta_1 + \cos \theta_2} \right|
$$

$$
\tag{2.92}
$$

As revealed by Eq. (2.92), unlike the maximum pseudo acceleration, the sliding displacement amplitude is dependent on the amplitude of the ground acceleration, a_g, for the sliding-sliding case. When $a_g/\mu g \to +\infty$, $\cos(\omega_g t_i) \to 0$ as revealed by Eq. (2.90), so the normalized sliding displacement amplitude, $u_{s,ap}/\left(a_g/\omega_g^2\right)$, in Eq. (2.92) tends toward to 2, which is consistent with the results presented in Sect. 2.5.2.

2.6 Conclusions

This chapter presents the responses of 2DOF sliding base systems subjected to harmonic ground motions. The response history of a SB system can exhibit two types of phases: the stick phase and the sliding phase. During the sliding phase, the vibration of the relative displacement has a higher natural frequency and damping ratio compared with the stick phase, which is related to the mass ratio. The equivalent dynamic force for the vibration of the relative displacement during the sliding phase is a step force, which results a static displacement as given in Eq. (2.12). The responses of the sliding base system under harmonic ground motions converge rapidly to steady periodic responses with the same period as the ground acceleration. The sliding base system displays three types of motions, namely stick-stick, stick-sliding, and sliding-sliding, as the ground acceleration amplitude increases.

A sliding base structure has multiple resonant frequencies when it slides, unlike fixed base structures. For the sliding-sliding case, resonance happens when the period of ground motion is odd times of the natural period of vibration of the relative displacement that occurs during sliding. As the mass ratio increases, the maximum pseudo acceleration generally decreases. However, Resonance can result in obtaining larger maximum pseudo acceleration for larger mass ratios in some particular cases. Moreover, in certain instances, the sliding base structure can have a higher maximum pseudo acceleration than its fixed base counterpart due to extra resonances. In the sliding-sliding case, the upper limit of pseudo acceleration is reached, regardless of the amplitude of the ground acceleration. Equations (2.80)–(2.83) can be used to calculate the maximum pseudo acceleration in the sliding-sliding case.

The amplitude of the sliding displacement is a suitable response quantity to represent the extent of sliding, and it can be normalized by the half of the vibration amplitude of the ground displacement, a_g/ω_g^2. The normalized sliding displacement amplitude, $u_{s,ap}/(a_g/\omega_g^2)$, is more affected by the mass ratio and frequency ratio, ω_g/ω, in region $\omega_g/\omega \geq 1$ compared to region $\omega_g/\omega < 1$. The normalized sliding displacement amplitude exhibits a higher value as $a_g/\mu g$ increases following the sliding occurring, yet its rate of increase gradually reduces, and it eventually reaches an upper bound of 2.

Chapter 3
Response Histories of Sliding Base Structures Under Earthquake Excitation

3.1 Equations of Motion

Figure 3.1a shows a schematic diagram of an N-story SB structure, the sliding base of which rests on the foundation. By lumping the structure mass at the corresponding floor level, the multistory SB structure can be analyzed using the simplified model shown in Fig. 3.1b. The lateral displacements of the ith floor in the x and y directions with respect to the sliding base are denoted as u_{rxi} and u_{ryi}, respectively. By further assuming that the structure is symmetrical about the x and y axes, the equations of dynamic equilibrium of the structure subjected to three-component excitations are as follows:

$$\begin{cases} \mathbf{m}\left[(\ddot{u}_{gx} + \ddot{u}_{sx})\mathbf{1} + \ddot{\mathbf{u}}_{rx}\right] + \mathbf{c}_x\dot{\mathbf{u}}_{rx} + \mathbf{k}_x\mathbf{u}_{rx} = \mathbf{0} \\ m_b(\ddot{u}_{gx} + \ddot{u}_{sx}) + \mathbf{1}^{\mathrm{T}}\mathbf{m}\left[(\ddot{u}_{gx} + \ddot{u}_{sx})\mathbf{1} + \ddot{\mathbf{u}}_{rx}\right] = f_x \\ \mathbf{m}\left[(\ddot{u}_{gy} + \ddot{u}_{sy})\mathbf{1} + \ddot{\mathbf{u}}_{ry}\right] + \mathbf{c}_y\dot{\mathbf{u}}_{ry} + \mathbf{k}_y\mathbf{u}_{ry} = \mathbf{0} \\ m_b(\ddot{u}_{gy} + \ddot{u}_{sy}) + \mathbf{1}^{\mathrm{T}}\mathbf{m}\left[(\ddot{u}_{gy} + \ddot{u}_{sy})\mathbf{1} + \ddot{\mathbf{u}}_{ry}\right] = f_y \end{cases} \tag{3.1}$$

where $\mathbf{m}$, $\mathbf{c}_x$ ($\mathbf{c}_y$), and $\mathbf{k}_x$ ($\mathbf{k}_y$) are the mass, damping, and stiffness matrices, respectively, when the structure base is fixed; m_b is the mass of the sliding base; $\mathbf{0}$ and $\mathbf{1}$ are the vectors whose elements are all zero and unity, respectively. $\ddot{u}_{gx}(t)$ and $\ddot{u}_{gy}(t)$ are the x and y components, respectively, of the ground acceleration; $\ddot{u}_{sx}(t)$ and $\ddot{u}_{sy}(t)$ are the sliding accelerations with respect to the ground in the x and y directions, respectively; $\mathbf{u}_{rx} = [u_{rx1}, u_{rx2}, \ldots, u_{rxN}]$ and $\mathbf{u}_{ry} = [u_{ry1}, u_{ry2}, \ldots, u_{ryN}]$ are the floor displacement vectors in the x and y directions, respectively; $\dot{\mathbf{u}}_{rx}$ and $\dot{\mathbf{u}}_{ry}$, and $\ddot{\mathbf{u}}_{rx}$ and $\ddot{\mathbf{u}}_{ry}$ are the corresponding velocity and acceleration vectors, respectively; and $f_x(t)$ and $f_y(t)$ are the x and y components, respectively, of the friction force at the sliding interface. In Eq. (3.1), the first and second (third and fourth) equations represent the dynamic equilibrium of each floor mass and the whole structure, respectively, in the x-direction (y-direction).

H.-S Hu, *Sliding Base Structures*,
https://doi.org/10.1007/978-981-99-5107-9_3

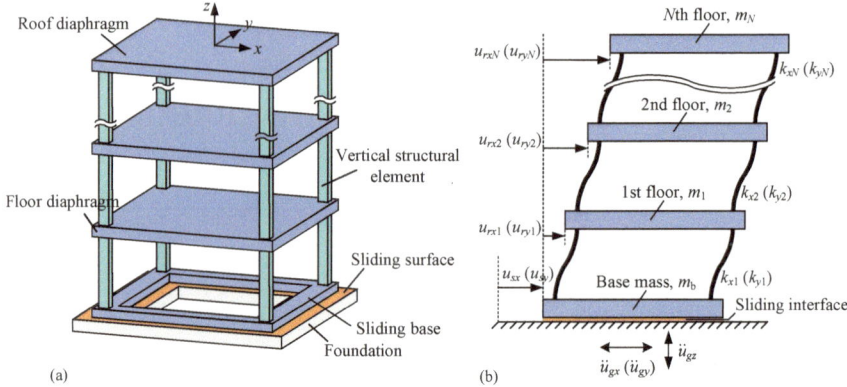

Fig. 3.1 *N*-story SB structure: **a** schematic diagram; and **b** simplified model

For the model shown in Fig. 3.1b, the mass matrix **m** is

$$\mathbf{m} = \begin{bmatrix} m_1 & 0 & 0 & \cdots & 0 \\ & m_2 & 0 & \cdots & 0 \\ & & m_3 & \cdots & 0 \\ & \text{sym.} & & \ddots & \vdots \\ & & & & m_N \end{bmatrix} \tag{3.2}$$

where m_i ($i = 1, 2, 3, …, N$) is the mass of the *i*th floor. The stiffness matrices $\mathbf{k}_x$ and $\mathbf{k}_y$ are

$$\mathbf{k}_x = \begin{bmatrix} k_{x1} + k_{x2} & -k_{x2} & 0 & \cdots & 0 & 0 \\ & k_{x2} + k_{x3} & -k_{x3} & \cdots & 0 & 0 \\ & & k_{x3} + k_{x4} & \cdots & 0 & 0 \\ & & & \ddots & \vdots & \vdots \\ & & & & -k_{x(N-1)} & 0 \\ & \text{sym.} & & & k_{x(N-1)} + k_{xN} & -k_{xN} \\ & & & & & k_{xN} \end{bmatrix} \tag{3.3}$$

and

$$\mathbf{k}_y = \begin{bmatrix} k_{y1} + k_{y2} & -k_{y2} & 0 & \cdots & 0 & 0 \\ & k_{y2} + k_{y3} & -k_{y3} & \cdots & 0 & 0 \\ & & k_{y3} + k_{y4} & \cdots & 0 & 0 \\ & & & \ddots & \vdots & \vdots \\ & & & & -k_{y(N-1)} & 0 \\ & \text{sym.} & & & k_{y(N-1)} + k_{yN} & -k_{yN} \\ & & & & & k_{yN} \end{bmatrix} \tag{3.4}$$

where k_{xi} and k_{yi} are the lateral stiffness of the ith story in the x and y directions, respectively. The damping of the superstructure can be assumed to be of the Rayleigh type; thus, when the damping ratios of two specified modes of the superstructure are given, the damping matrix $\mathbf{c}_x$ and $\mathbf{c}_y$ can then be determined.

The response history of an SB structure can exhibit two types of phases: stick phase and sliding phase. For the stick phases, during which sliding does not occur, the sliding acceleration is equal to 0, and the friction force, f, is smaller than the static friction force; thus, we have

$$\begin{cases} \ddot{u}_{sx} = \ddot{u}_{sy} = 0 \\ f = \sqrt{f_x^2 + f_y^2} < (m + m_b)(g + \ddot{u}_{gz})\mu_s \end{cases} \tag{3.5}$$

where $\ddot{u}_{gz}(t)$ is the z (vertical) component of the ground acceleration, g is the gravity acceleration, and μ_s is the static friction coefficient. Combining Eqs. (3.1) and (3.5) leads to

$$\begin{cases} \mathbf{m}[\ddot{u}_{gx}\mathbf{1} + \ddot{\mathbf{u}}_{rx}] + \mathbf{c}_x \dot{\mathbf{u}}_{rx} + \mathbf{k}_x \mathbf{u}_{rx} = \mathbf{0} \\ \mathbf{m}[\ddot{u}_{gy}\mathbf{1} + \ddot{\mathbf{u}}_{ry}] + \mathbf{c}_y \dot{\mathbf{u}}_{ry} + \mathbf{k}_y \mathbf{u}_{ry} = \mathbf{0} \\ \sqrt{\left(\dfrac{\mathbf{1}^\mathrm{T}\mathbf{m}\ddot{\mathbf{u}}_{rx}}{\sum_{i=1}^{N} m_i + m_b} + \ddot{u}_{gx} \right)^2 + \left(\dfrac{\mathbf{1}^\mathrm{T}\mathbf{m}\ddot{\mathbf{u}}_{rx}}{\sum_{i=1}^{N} m_i + m_b} + \ddot{u}_{gy} \right)^2} < (g + \ddot{u}_{gz})\mu_s \end{cases} \tag{3.6}$$

The first and second equations of Eq. (3.6) govern the response of an SB structure during the stick phases. The third equation of Eq. (3.6) is the precondition for the stick phases; when it is no longer satisfied, a sliding phase starts.

During the sliding phases, the direction of the friction force is opposite to the direction of the sliding velocity; thus, we have

$$\begin{cases} f_x = \dfrac{-\dot{u}_{sx}}{\sqrt{\dot{u}_{sx}^2 + \dot{u}_{sy}^2}} \left(\displaystyle\sum_{i=1}^{N} m_i + m_b \right) \left(g + \ddot{u}_{gz} \right) \mu \\[3em] f_y = \dfrac{-\dot{u}_{sy}}{\sqrt{\dot{u}_{sx}^2 + \dot{u}_{sy}^2}} \left(\displaystyle\sum_{i=1}^{N} m_i + m_b \right) \left(g + \ddot{u}_{gz} \right) \mu \end{cases} \tag{3.7}$$

where μ is the dynamic friction coefficient. Combining Eqs. (3.1) and (3.7) yields

$$\begin{cases} \ddot{u}_{sx}\mathbf{m1} + \mathbf{m}\ddot{\mathbf{u}}_{rx} + c_x \dot{\mathbf{u}}_{rx} + k_x \mathbf{u}_{rx} = -\ddot{u}_{gx}\mathbf{m1} \\[1em] \ddot{u}_{sx} + \dfrac{\mathbf{1}^{\mathrm{T}}\mathbf{m}\ddot{\mathbf{u}}_{rx}}{\sum_{i=1}^{N} m_i + m_b} + \dfrac{\dot{u}_{sx}}{\sqrt{\dot{u}_{sx}^2 + \dot{u}_{sy}^2}} \left(g + \ddot{u}_{gz} \right) \mu = -\ddot{u}_{gx} \\[2em] \ddot{u}_{sy}\mathbf{m1} + \mathbf{m}\ddot{\mathbf{u}}_{ry} + c_y \dot{\mathbf{u}}_{ry} + k_y \mathbf{u}_{ry} = -\ddot{u}_{gy}\mathbf{m1} \\[1em] \ddot{u}_{sy} + \dfrac{\mathbf{1}^{\mathrm{T}}\mathbf{m}\ddot{\mathbf{u}}_{ry}}{\sum_{i=1}^{N} m_i + m_b} + \dfrac{\dot{u}_{sy}}{\sqrt{\dot{u}_{sx}^2 + \dot{u}_{sy}^2}} \left(g + \ddot{u}_{gz} \right) \mu = -\ddot{u}_{gy} \end{cases} \tag{3.8}$$

Equation (3.8) governs the response of an SB structure during the sliding phases. When the sliding velocity during a sliding phase becomes 0, this round of sliding ends. Afterwards, the structure may continue to slide or enter a stick phase depending on whether the third equation of Eq. (3.6) is satisfied. Based on the governing equations for the stick and sliding phases and the transition conditions between different phases, as presented above, a program was developed for computing the responses of SB structures subjected to three-dimensional excitations. The numerical methods for solving Eq. (3.8) will be presented in the next section.

3.2 Numerical Computation Methods

The first and second equations of Eq. (3.6), which govern the response of an SB structure during the stick phases, are the same as the governing equations of a fixed base structure. The numerical methods for solving these equations can be found in Chopra (2001).

The system of differential equations presented in Eq. (3.8) can be solved by the time-stepping method. The response quantities $(\ddot{u}_{sx})_{i+1}$, $(\ddot{u}_{sy})_{i+1}$, $(\mathbf{u}_{rx})_{i+1}$, $(\dot{\mathbf{u}}_{rx})_{i+1}$, $(\ddot{\mathbf{u}}_{rx})_{i+1}$, $(\mathbf{u}_{ry})_{i+1}$, $(\dot{\mathbf{u}}_{ry})_{i+1}$ and $(\ddot{\mathbf{u}}_{ry})_{i+1}$ at time $i+1$ satisfy Eq. (3.8) at time $i+1$, i.e.,

$$
\begin{cases}
(\ddot{u}_{sx})_{i+1}\mathbf{m1} + \mathbf{m}(\ddot{\mathbf{u}}_{rx})_{i+1} + \mathbf{c}_x(\dot{\mathbf{u}}_{rx})_{i+1} + \mathbf{k}_x(\mathbf{u}_{rx})_{i+1} = -(\ddot{u}_{gx})_{i+1}\mathbf{m1} \\[2mm]
(\ddot{u}_{sx})_{i+1} + \dfrac{\mathbf{1}^{\mathrm{T}}\mathbf{m}(\ddot{\mathbf{u}}_{rx})_{i+1}}{\sum_{i=1}^{N} m_i + m_b} + \dfrac{(\dot{u}_{sx})_{i+1}}{\sqrt{\left[(\dot{u}_{sx})_{i+1}\right]^2 + \left[(\dot{u}_{sy})_{i+1}\right]^2}} \\[4mm]
\qquad \left[g + (\ddot{u}_{gz})_{i+1}\right]\mu = -(\ddot{u}_{gx})_{i+1} \\[2mm]
(\ddot{u}_{sy})_{i+1}\mathbf{m1} + \mathbf{m}(\ddot{\mathbf{u}}_{ry})_{i+1} + \mathbf{c}_y(\dot{\mathbf{u}}_{ry})_{i+1} + \mathbf{k}_y(\mathbf{u}_{ry})_{i+1} = -(\ddot{u}_{gy})_{i+1}\mathbf{m1} \\[2mm]
(\ddot{u}_{sy})_{i+1} + \dfrac{\mathbf{1}^{\mathrm{T}}\mathbf{m}(\ddot{\mathbf{u}}_{ry})_{i+1}}{\sum_{i=1}^{N} m_i + m_b} + \dfrac{(\dot{u}_{sy})_{i+1}}{\sqrt{\left[(\dot{u}_{sx})_{i+1}\right]^2 + \left[(\dot{u}_{sy})_{i+1}\right]^2}} \\[4mm]
\qquad (g + \ddot{u}_{gz})\mu = -(\ddot{u}_{gy})_{i+1}
\end{cases}
\tag{3.9}
$$

Using Newmark's equations (Newmark, 1959) for the relationships between the response quantities at time $i+1$ and the corresponding quantities at time i, we have

$$
\begin{cases}
(\dot{u}_{sx})_{i+1} = (\dot{u}_{sx})_i + \dfrac{1}{2}\Delta t(\ddot{u}_{sx})_i + \dfrac{1}{2}\Delta t(\ddot{u}_{sx})_{i+1} \\[2mm]
(\mathbf{u}_{rx})_{i+1} = (\mathbf{u}_{rx})_i + \Delta t(\dot{\mathbf{u}}_{rx})_i + [(0.5 - \beta)(\Delta t)](\ddot{\mathbf{u}}_{rx})_i + \left[\beta(\Delta t)^2\right](\ddot{\mathbf{u}}_{rx})_{i+1} \\[2mm]
(\dot{\mathbf{u}}_{rx})_{i+1} = (\dot{\mathbf{u}}_{rx})_i + \dfrac{1}{2}\Delta t(\ddot{\mathbf{u}}_{rx})_i + \dfrac{1}{2}\Delta t(\ddot{\mathbf{u}}_{rx})_{i+1} \\[2mm]
(\dot{u}_{sy})_{i+1} = (\dot{u}_{sy})_i + \dfrac{1}{2}\Delta t(\ddot{u}_{sy})_i + \dfrac{1}{2}\Delta t(\ddot{u}_{sy})_{i+1} \\[2mm]
(\mathbf{u}_{ry})_{i+1} = (\mathbf{u}_{ry})_i + \Delta t(\dot{\mathbf{u}}_{ry})_i + [(0.5 - \beta)(\Delta t)](\ddot{\mathbf{u}}_{ry})_i + \left[\beta(\Delta t)^2\right](\ddot{\mathbf{u}}_{ry})_{i+1} \\[2mm]
(\dot{\mathbf{u}}_{ry})_{i+1} = (\dot{\mathbf{u}}_{ry})_i + \dfrac{1}{2}\Delta t(\ddot{\mathbf{u}}_{ry})_i + \dfrac{1}{2}\Delta t(\ddot{\mathbf{u}}_{ry})_{i+1}
\end{cases}
\tag{3.10}
$$

where Δt is the time step, and β, which ranges from 1/6 to 1/4, is a parameter that controls the variation in the acceleration over a time step. Equation (3.10) can be converted to

$$
\begin{cases}
(\ddot{u}_{sx})_{i+1} = \dfrac{2}{\Delta t}\left[(\dot{u}_{sx})_{i+1} - (\dot{u}_{sx})_i\right] - (\ddot{u}_{sx})_i \\[2mm]
(\ddot{\mathbf{u}}_{rx})_{i+1} = \dfrac{1}{\beta(\Delta t)^2}\left[(\mathbf{u}_{rx})_{i+1} - (\mathbf{u}_{rx})_i\right] - \dfrac{1}{\beta \Delta t}(\dot{\mathbf{u}}_{rx})_i - \left(\dfrac{1}{2\beta} - 1\right)(\ddot{\mathbf{u}}_{rx})_i \\[2mm]
(\dot{\mathbf{u}}_{rx})_{i+1} = \dfrac{1}{2\beta \Delta t}\left[(\mathbf{u}_{rx})_{i+1} - (\mathbf{u}_{rx})_i\right] + \left(1 - \dfrac{1}{2\beta}\right)(\dot{\mathbf{u}}_{rx})_i + \left(\Delta t - \dfrac{\Delta t}{4\beta}\right)(\ddot{\mathbf{u}}_{rx})_i \\[2mm]
(\ddot{u}_{sy})_{i+1} = \dfrac{2}{\Delta t}\left[(\dot{u}_{sy})_{i+1} - (\dot{u}_{sy})_i\right] - (\ddot{u}_{sy})_i \\[2mm]
(\ddot{\mathbf{u}}_{ry})_{i+1} = \dfrac{1}{\beta(\Delta t)^2}\left[(\mathbf{u}_{ry})_{i+1} - (\mathbf{u}_{ry})_i\right] - \dfrac{1}{\beta \Delta t}(\dot{\mathbf{u}}_{ry})_i - \left(\dfrac{1}{2\beta} - 1\right)(\ddot{\mathbf{u}}_{ry})_i \\[2mm]
(\dot{\mathbf{u}}_{ry})_{i+1} = \dfrac{1}{2\beta \Delta t}\left[(\mathbf{u}_{ry})_{i+1} - (\mathbf{u}_{ry})_i\right] + \left(1 - \dfrac{1}{2\beta}\right)(\dot{\mathbf{u}}_{ry})_i + \left(\Delta t - \dfrac{\Delta t}{4\beta}\right)(\ddot{\mathbf{u}}_{ry})_i
\end{cases}
\tag{3.11}
$$

Substituting Eq. (3.11) into Eq. (3.9) gives

$$
\begin{cases}
\dfrac{2}{\Delta t}(\dot{u}_{sx})_{i+1}\mathbf{m1} + \left[\dfrac{1}{\beta(\Delta t)^2}\mathbf{m} + \dfrac{1}{2\beta \Delta t}\mathbf{c}_x + \mathbf{k}_x\right](\mathbf{u}_{rx})_{i+1} = \mathbf{p}_1 \\[3mm]
\dfrac{2}{\Delta t}(\dot{u}_{sx})_{i+1} + \dfrac{(\dot{u}_{sx})_{i+1}}{\sqrt{\left[(\dot{u}_{sx})_{i+1}\right]^2 + \left[(\dot{u}_{sy})_{i+1}\right]^2}}\left[g + (\ddot{u}_{gz})_{i+1}\right]\mu \\[3mm]
\qquad\qquad + \dfrac{1}{\beta(\Delta t)^2}\dfrac{\mathbf{1}^{\mathrm{T}}\mathbf{m}(\mathbf{u}_{rx})_{i+1}}{\sum_{i=1}^{N} m_i + m_b} = p_2 \\[3mm]
\dfrac{2}{\Delta t}(\dot{u}_{sy})_{i+1}\mathbf{m1} + \left[\dfrac{1}{\beta(\Delta t)^2}\mathbf{m} + \dfrac{1}{2\beta \Delta t}\mathbf{c}_y + \mathbf{k}_y\right](\mathbf{u}_{ry})_{i+1} = \mathbf{p}_3 \\[3mm]
\dfrac{2}{\Delta t}(\dot{u}_{sy})_{i+1} + \dfrac{(\dot{u}_{sy})_{i+1}}{\sqrt{\left[(\dot{u}_{sx})_{i+1}\right]^2 + \left[(\dot{u}_{sy})_{i+1}\right]^2}}\left[g + (\ddot{u}_{gz})_{i+1}\right]\mu \\[3mm]
\qquad\qquad + \dfrac{1}{\beta(\Delta t)^2}\dfrac{\mathbf{1}^{\mathrm{T}}\mathbf{m}(\mathbf{u}_{ry})_{i+1}}{\sum_{i=1}^{N} m_i + m_b} = p_4
\end{cases}
\tag{3.12}
$$

where

$$
\begin{cases}
\mathbf{p}_1 = \left[-(\ddot{u}_{gx})_{i+1} + \dfrac{2}{\Delta t}(\dot{u}_{sx})_i + (\ddot{u}_{sx})_i \right]\mathbf{m1} + \left[\dfrac{1}{\beta(\Delta t)^2}\mathbf{m} + \dfrac{1}{2\beta\Delta t}\mathbf{c}_x \right](\mathbf{u}_{rx})_i \\[4mm]
\qquad + \left[\dfrac{1}{\beta\Delta t}\mathbf{m} - \left(1 - \dfrac{1}{2\beta}\right)\mathbf{c}_x \right](\dot{\mathbf{u}}_{rx})_i + \left[\left(\dfrac{1}{2\beta} - 1\right)\mathbf{m} - \left(\Delta t - \dfrac{\Delta t}{4\beta}\right)\mathbf{c}_x \right](\ddot{\mathbf{u}}_{rx})_i \\[4mm]
p_2 = -(\ddot{u}_{gx})_{i+1} + \dfrac{2}{\Delta t}(\dot{u}_{sx})_i + (\ddot{u}_{sx})_i \\[4mm]
\qquad + \dfrac{\mathbf{1}^{\mathrm{T}}\mathbf{m}}{\sum_{i=1}^{N} m_i + m_b}\left[\dfrac{1}{\beta(\Delta t)^2}(\mathbf{u}_{rx})_i + \dfrac{1}{\beta\Delta t}(\dot{u}_{rx})_i + \left(\dfrac{1}{2\beta} - 1\right)(\ddot{u}_{rx})_i \right] \\[4mm]
\mathbf{p}_3 = \left[-(\ddot{u}_{gy})_{i+1} + \dfrac{2}{\Delta t}(\dot{u}_{sy})_i + (\ddot{u}_{sy})_i \right]\mathbf{m1} + \left[\dfrac{1}{\beta(\Delta t)^2}\mathbf{m} + \dfrac{1}{2\beta\Delta t}\mathbf{c}_y \right](\mathbf{u}_{ry})_i \\[4mm]
\qquad + \left[\dfrac{1}{\beta\Delta t}\mathbf{m} - \left(1 - \dfrac{1}{2\beta}\right)\mathbf{c}_y \right](\dot{\mathbf{u}}_{ry})_i + \left[\left(\dfrac{1}{2\beta} - 1\right)\mathbf{m} - \left(\Delta t - \dfrac{\Delta t}{4\beta}\right)\mathbf{c}_y \right](\ddot{\mathbf{u}}_{ry})_i \\[4mm]
p_4 = -(\ddot{u}_{gy})_{i+1} + \dfrac{2}{\Delta t}(\dot{u}_{sy})_i + (\ddot{u}_{sy})_i \\[4mm]
\qquad + \dfrac{\mathbf{1}^{\mathrm{T}}\mathbf{m}}{\sum_{i=1}^{N} m_i + m_b}\left[\dfrac{1}{\beta(\Delta t)^2}(\mathbf{u}_{ry})_i + \dfrac{1}{\beta\Delta t}(\dot{u}_{ry})_i + \left(\dfrac{1}{2\beta} - 1\right)(\ddot{u}_{ry})_i \right]
\end{cases}
\tag{3.13}
$$

When determining the response quantities at time $i + 1$, the response quantities at time i are already known; thus, Eq. (3.12) is a system of nonlinear equations with four unknowns, namely, $(\dot{u}_{sx})_{i+1}$, $(\mathbf{u}_{rx})_{i+1}$, $(\dot{u}_{sy})_{i+1}$ and $(\mathbf{u}_{ry})_{i+1}$. These nonlinear equations can be solved using the Newton–Raphson iteration technique (Chopra, 2001). After determining the values of $(\dot{u}_{sx})_{i+1}$, $(\mathbf{u}_{rx})_{i+1}$, $(\dot{u}_{sy})_{i+1}$ and $(\mathbf{u}_{ry})_{i+1}$, other response quantities, namely, $(u_{sx})_{i+1}$, $(\ddot{u}_{sx})_{i+1}$, $(\dot{\mathbf{u}}_{rx})_{i+1}$, $(\ddot{\mathbf{u}}_{rx})_{i+1}$, $(u_{sy})_{i+1}$, $(\ddot{u}_{sy})_{i+1}$, $(\dot{\mathbf{u}}_{ry})_{i+1}$ and $(\ddot{\mathbf{u}}_{ry})_{i+1}$ at time $i + 1$ can also be determined using Newmark's equations.

3.3 Response Histories Under Earthquake Excitation

Figure 3.2 shows the x-component of the ground accelerations recorded at the Mammoth Lakes station during the 1980 Mammoth Lakes earthquake. The peak ground acceleration (PGA) in the x-direction is 3.8 m/s². Figure 3.3 shows the absolute acceleration history of the top mass of a single-story fixed base structure subjected to this ground motion. The natural periods of the superstructure along the two principal axes are both 0.3 s, and the damping ratios are taken as 5%. The peak absolute acceleration of the top mass is 7.66 m/s², which is approximately 2 times the corresponding PGA. Figure 3.4 shows the responses of SB structures with different friction coefficients subjected to this ground motion. The superstructure is the same

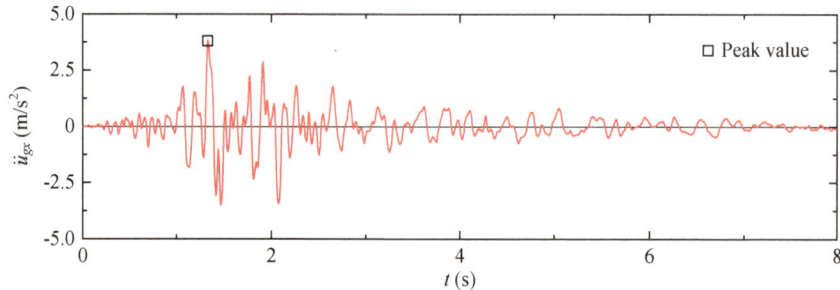

Fig. 3.2 *x*-component of the ground accelerations recorded at the Mammoth Lakes station during the 1980 Mammoth Lakes earthquake

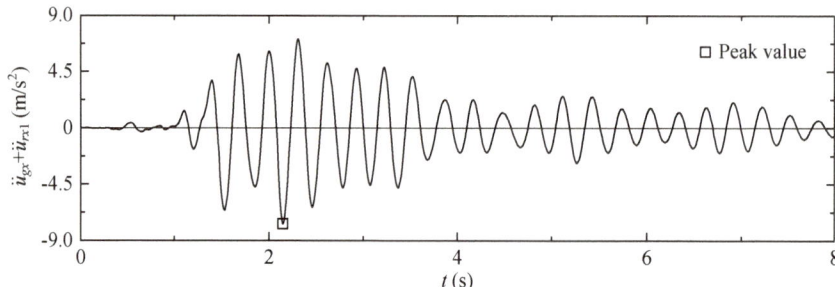

Fig. 3.3 Absolute acceleration history of the top mass of a single-story fixed base structure subjected to the Mammoth Lakes record

as that of the aforementioned fixed base structure. The mass of the sliding base is 3/7 of the top mass. As shown in Fig. 3.4, the peak absolute acceleration of the top mass decreases quickly as the friction coefficient decreases; when $\mu = \mu_s = 0.05$, the peak absolute acceleration of the top mass is 1.16 m/s^2, which is only 15% of the peak absolute acceleration when the base is fixed. The response history of the sliding displacement varies significantly when different friction coefficients are used. As the friction coefficient decreases, sliding occurs more frequently. Furthermore, the maximum sliding displacements may be obtained in the opposite directions for different friction coefficients.

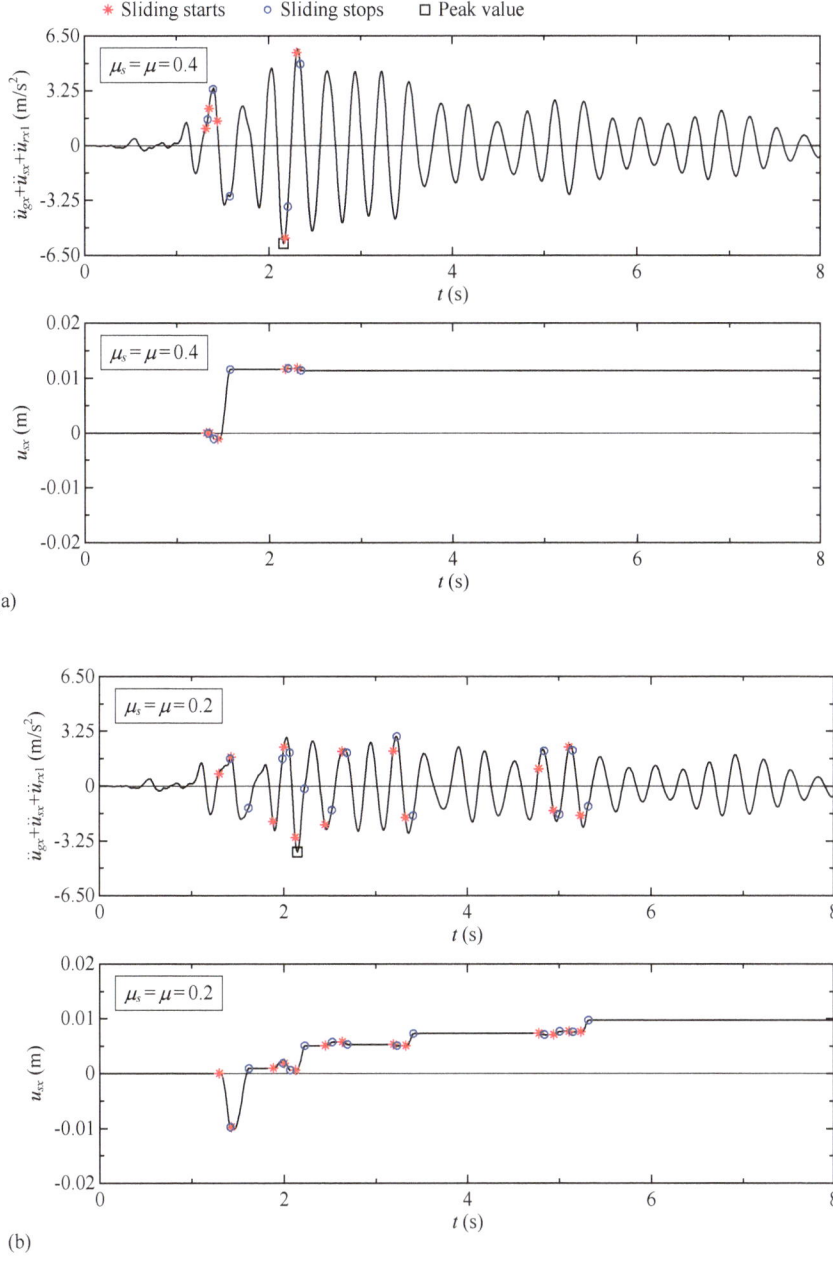

Fig. 3.4 Response histories of SB structures with different friction coefficients subjected to the Mammoth Lakes record

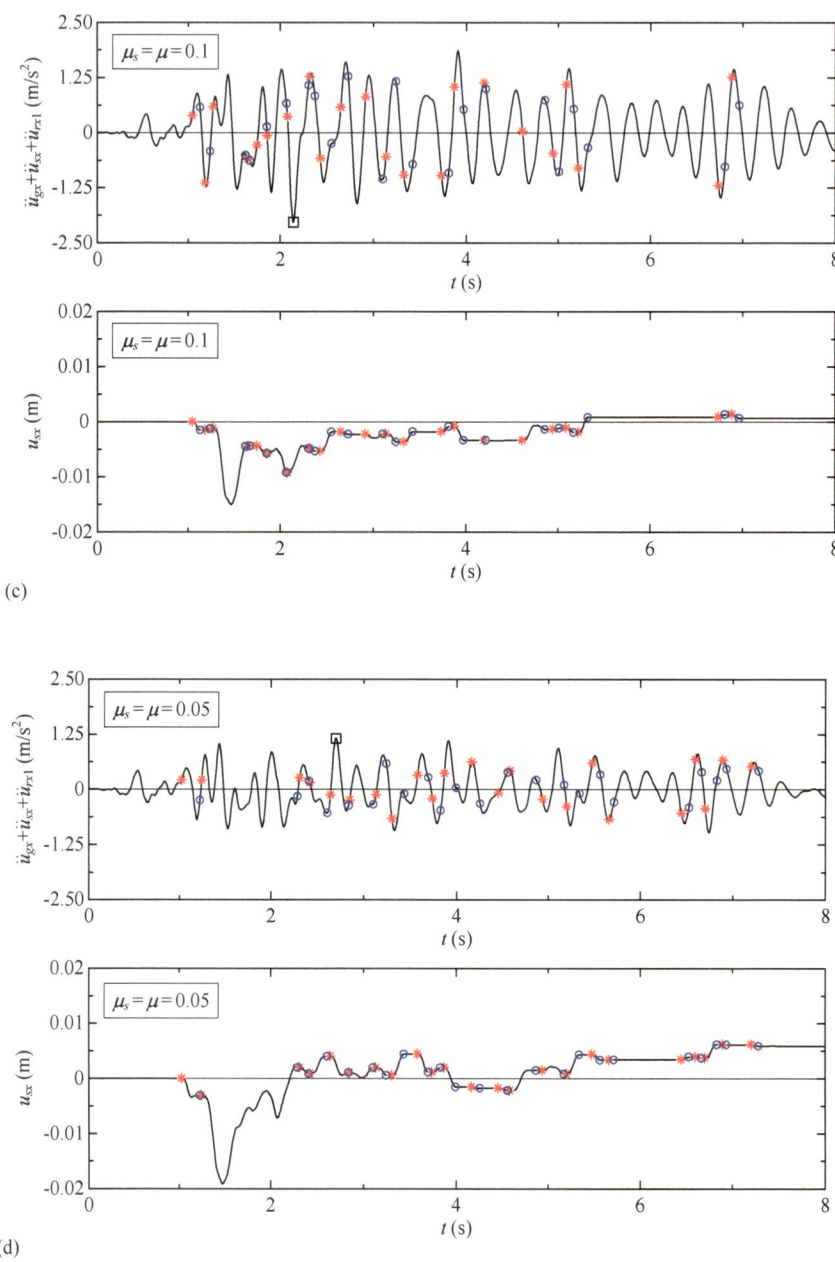

(c)

(d)

Fig. 3.4 (continued)

Chapter 4
Peak Superstructure Responses of Single-Story Sliding Base Structures Under Earthquake Excitation

4.1 Critical Parameters and Their Ranges

In Chap. 3, the equations of motion for multistory SB structures have been derived. For single-story SB structures (Fig. 4.1), the governing equations can be simplified to the following forms:

$$\begin{cases} \ddot{u}_{rx} + 2\xi_x \omega_x \dot{u}_{rx} + \omega_x^2 u_{rx} = -\ddot{u}_{gx} \\ \ddot{u}_{ry} + 2\xi_y \omega_y \dot{u}_{ry} + \omega_y^2 u_{ry} = -\ddot{u}_{gy} \\ \sqrt{\left(\alpha\ddot{u}_{rx} + \ddot{u}_{gx}\right)^2 + \left(\alpha\ddot{u}_{ry} + \ddot{u}_{gy}\right)^2} < \left(g + \ddot{u}_{gz}\right)\mu_s \end{cases} \tag{4.1}$$

for the stick phases, and

$$\begin{cases} \ddot{u}_{sx} + \ddot{u}_{rx} + 2\xi_x \omega_x \dot{u}_{rx} + \omega_x^2 u_{rx} = -\ddot{u}_{gx} \\ \ddot{u}_{sx} + \dfrac{\dot{u}_{sx}}{\sqrt{\dot{u}_{sx}^2 + \dot{u}_{sy}^2}}\left(g + \ddot{u}_{gz}\right)\mu + \alpha\ddot{u}_{rx} = -\ddot{u}_{gx} \\ \ddot{u}_{sy} + \ddot{u}_{ry} + 2\xi_y \omega_y \dot{u}_{ry} + \omega_y^2 u_{ry} = -\ddot{u}_{gy} \\ \ddot{u}_{sy} + \dfrac{\dot{u}_{sy}}{\sqrt{\dot{u}_{sx}^2 + \dot{u}_{sy}^2}}\left(g + \ddot{u}_{gz}\right)\mu + \alpha\ddot{u}_{ry} = -\ddot{u}_{gy} \end{cases} \tag{4.2}$$

for the sliding phases. In Eqs. (4.1) and (4.2), $\omega_x = \sqrt{k_x/m}$ and $\xi_x = c_x/(2m\omega_x)$ ($\omega_y = \sqrt{k_y/m}$ and $\xi_y = c_y/(2m\omega_y)$) are the natural frequency and damping ratio, respectively, of the corresponding fixed base (FB) structure in the x-direction (y-direction), and $\alpha = m/(m + m_b)$ is defined as the mass ratio.

As revealed by Eqs. (4.1) and (4.2), the response of an SB structure is greatly affected by the friction coefficient; thus, it is meaningful to relate the intensity of the ground motion and the acceleration response of the superstructure to the friction

H.-S Hu, *Sliding Base Structures*, https://doi.org/10.1007/978-981-99-5107-9_4

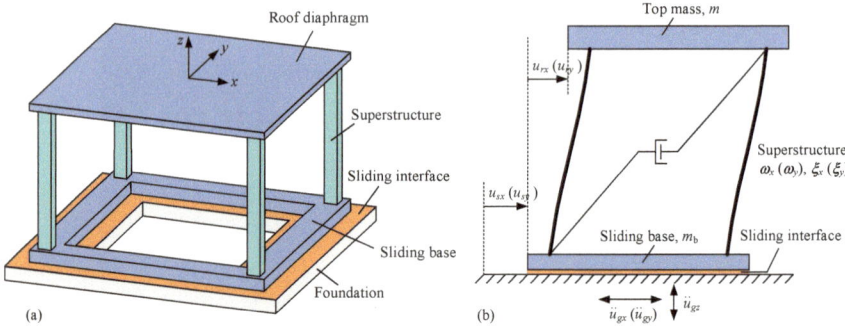

Fig. 4.1 Single-story SB structure: **a** schematic plot; and **b** simplified model

coefficient, μ. To consider the response in the x-direction, we introduce

$$u^*_{stx} = \frac{\mu g}{\omega_x^2} \tag{4.3}$$

which is the displacement of the corresponding FB structure when subjected to a static force, $mg\mu$, in the x-direction. Dividing Eqs. (4.1) and (4.2) by u^*_{stx} leads to the following equations:

$$
\begin{cases}
\ddot{\bar{u}}_{rx} + 2\xi_x\omega_x\dot{\bar{u}}_{rx} + \omega_x^2\bar{u}_{rx} = -\omega_x^2\dfrac{a_{gx0}}{\mu g}\ddot{\bar{u}}_{gx} \\[2ex]
\ddot{\bar{u}}_{ry} + 2\xi_y\omega_x\left(\dfrac{\omega_y}{\omega_x}\right)\dot{\bar{u}}_{ry} + \omega_x^2\left(\dfrac{\omega_y}{\omega_x}\right)^2\bar{u}_{ry} = -\omega_x^2\dfrac{a_{gx0}}{\mu g}\dfrac{a_{gy0}}{a_{gx0}}\ddot{\bar{u}}_{gy} \\[2ex]
\sqrt{\left(\alpha\ddot{\bar{u}}_{rx} + \dfrac{a_{gx0}}{\mu g}\ddot{\bar{u}}_{gx}\omega_x^2\right)^2 + \left(\alpha\ddot{\bar{u}}_{ry} + \dfrac{a_{gy0}}{\mu g}\ddot{\bar{u}}_{gy}\omega_x^2\right)^2} < \left(1 + \dfrac{\ddot{u}_{gz}}{g}\right)\omega_x^2\dfrac{\mu_s}{\mu}
\end{cases} \tag{4.4}
$$

and

$$
\begin{cases}
\ddot{\bar{u}}_{sx} + \ddot{\bar{u}}_{rx} + 2\xi_x\omega_x\dot{\bar{u}}_{rx} + \omega_x^2\bar{u}_{rx} = -\omega_x^2\dfrac{a_{gx0}}{\mu g}\ddot{\bar{u}}_{gx} \\[2ex]
\ddot{\bar{u}}_{sx} + \dfrac{\dot{\bar{u}}_{sx}}{\sqrt{\dot{\bar{u}}_{sx}^2 + \dot{\bar{u}}_{sy}^2}}\left(1 + \dfrac{\ddot{u}_{gz}}{g}\right)\omega_x^2 + \alpha\ddot{\bar{u}}_{rx} = -\omega_x^2\dfrac{a_{gx0}}{\mu g}\ddot{\bar{u}}_{gx} \\[2ex]
\ddot{\bar{u}}_{sy} + \ddot{\bar{u}}_{ry} + 2\xi_y\omega_x\left(\dfrac{\omega_y}{\omega_x}\right)\dot{\bar{u}}_{ry} + \omega_x^2\left(\dfrac{\omega_y}{\omega_x}\right)^2\bar{u}_{ry} = -\omega_x^2\dfrac{a_{gx0}}{\mu g}\dfrac{a_{gy0}}{a_{gx0}}\ddot{\bar{u}}_{gy} \\[2ex]
\ddot{\bar{u}}_{sy} + \dfrac{\dot{\bar{u}}_{sy}}{\sqrt{\dot{\bar{u}}_{sx}^2 + \dot{\bar{u}}_{sy}^2}}\left(1 + \dfrac{\ddot{u}_{gz}}{g}\right)\omega_x^2\left(\dfrac{\omega_y}{\omega_x}\right)^2 + \alpha\ddot{\bar{u}}_{ry} = -\omega_x^2\dfrac{a_{gx0}}{\mu g}\dfrac{a_{gy0}}{a_{gx0}}\ddot{\bar{u}}_{gy}
\end{cases} \tag{4.5}
$$

where $\bar{u}_{rx}(t) = u_{rx}(t)/u^*_{stx}$ and $\bar{u}_{ry}(t) = u_{ry}(t)/u^*_{stx}$ are the normalized relative displacements and $\bar{u}_{sx}(t) = u_{sx}(t)/u^*_{stx}$ and $\bar{u}_{sy}(t) = u_{sy}(t)/u^*_{stx}$ are the normalized sliding displacements; $\dot{\bar{u}}_{rx}(t), \dot{\bar{u}}_{ry}(t), \dot{\bar{u}}_{sx}(t)$ and $\dot{\bar{u}}_{sy}(t)$ and $\ddot{\bar{u}}_{rx}(t), \ddot{\bar{u}}_{ry}(t), \ddot{\bar{u}}_{sx}(t)$ and $\ddot{\bar{u}}_{sy}(t)$ are the corresponding normalized velocities and accelerations, respectively; a_{gx0} and a_{gy0} are the peak values of the x and y components, respectively, of the ground acceleration; and $\ddot{\bar{u}}_{gx}(t) = \ddot{u}_{gx}(t)/a_{gx0}$ and $\ddot{\bar{u}}_{gy}(t) = \ddot{u}_{gy}(t)/a_{gy0}$ represent the waveforms of the ground acceleration history. Equations (4.4) and (4.5) imply that the normalized displacements, namely, $\bar{u}_{rx}(t), \bar{u}_{ry}(t), \bar{u}_{sx}(t)$ and $\bar{u}_{sy}(t)$, are dependent only on $\omega_x, \omega_y/\omega_x, \xi_x, \xi_y, \alpha, \mu_s/\mu, a_{gx0}/\mu g, a_{gy0}/a_{gx0}, \ddot{\bar{u}}_{gx}(t), \ddot{\bar{u}}_{gy}(t)$ and $\ddot{u}_{gz}(t)/g$, among which $a_{gy0}/a_{gx0}, \ddot{\bar{u}}_{gx}(t), \ddot{\bar{u}}_{gy}(t)$ and $\ddot{u}_{gz}(t)/g$ are associated with the ground motion characteristics.

The maximum earthquake force applied to the superstructure in the x-direction can be expressed as mA_x, where A_x is the peak pseudoacceleration (Chopra, 2001) in the x-direction, which is defined as

$$A_x = \omega_x^2 \times \max(|u_{rx}(t)|) \tag{4.6}$$

By using Eq. (4.3), the normalized peak pseudoacceleration, $A_x/\mu g$, can be written as

$$\frac{A_x}{\mu g} = \frac{\omega_x^2 \times \max(|u_{rx}(t)|)}{\mu g} = \frac{\max(|u_{rx}(t)|)}{u^*_{stx}} = \max(|\bar{u}_{rx}(t)|) \tag{4.7}$$

Therefore, $A_x/\mu g$ is equivalent to the maximum absolute value of the normalized relative displacement, $\bar{u}_{rx}(t)$. Since $\bar{u}_{rx}(t)$ is dependent only on $\omega_x, \omega_y/\omega_x, \xi_x, \xi_y, \alpha, \mu_s/\mu, a_{gx0}/\mu g, a_{gy0}/a_{gx0}, \ddot{\bar{u}}_{gx}(t), \ddot{\bar{u}}_{gy}(t)$ and $\ddot{u}_{gz}(t)/g, A_x/\mu g$ is also only dependent on these parameters. According to the principle of symmetry, the parameters that determine the response in the y-direction are the same as those that determine the response in the x-direction.

The first step towards conducting parametric studies in the following sections is to investigate the ranges of the critical parameters in accordance with common practice. Due to the fact that SB structures are designed for low-rise buildings, the natural periods of the superstructure, denoted as $T_x = 2\pi/\omega_x$ and $T_y = 2\pi/\omega_y$ in the x-direction and y-direction, respectively, are limited under 1.0 s. Building structures are typically designed with similar stiffnesses in two orthogonal directions; therefore, the range of $T_x/T_y = \omega_y/\omega_x$, the ratio of superstructure periods in the two horizontal directions, is assumed to be from 1/2 to 2. For the general application of SB isolation in masonry structures (e.g., Nanda et al., 2015; Qamaruddin et al., 1986), bond beams are constructed under the masonry walls as the SB element; since the bond beams weigh less than the roof (or floor) diaphragm, the resulting mass ratio, α, will be larger than 0.5 (Qamaruddin et al., 1986). When using sliding isolation bearings (e.g., Jampole et al., 2016), the mass of every floor is considered nearly equal; thus, the resulting mass ratio is approximately 0.5 for single-story buildings and over 0.5 for multistory buildings. Based on the statements provided, the mass ratio, α, is taken to be not less than 0.5 for subsequent analyses. Several studies (Barbagallo et al., 2017; Nanda et al., 2012, 2015; Yegian et al., 2004) investigating friction characteristics of sliding interfaces have found that the static friction coefficient, μ_s, is slightly greater than the dynamic friction coefficient μ; previous studies (Yegian et al., 2004) found that the maximum observed value of μ_s/μ was 1.38. Therefore, μ_s/μ is taken to range from 1.0 to 1.4. The damping ratios in the x and y directions (i.e., ξ_x and ξ_y, respectively) are both taken as 5%, which is a commonly adopted value. The range of the values of the dynamic friction coefficient, μ, of the sliding interfaces investigated for SB structures (Barbagallo et al., 2017; Hasani, 1996; Jampole et al., 2016; Nanda et al., 2012; Qamaruddin et al., 1986; Yegian et al., 2004) is from 0.07 to 0.41. Furthermore, since the peak ground acceleration (PGA) seldom exceeds 1.2g, the maximum value considered for $a_{gx0}/\mu g$ is taken as 20, i.e., $a_{gx0} = 1.4g$ if $\mu = 0.07$.

4.2   Earthquake Ground Motions Considered

The ground motion records were selected for each site class defined by ASCE 7-10 (ASCE, 2010) from the Pacific Earthquake Engineering Research Center-Next Generation Attenuation (PEER-NGA) database. Since SB isolation takes effect primarily under significant ground accelerations, it is advisable to select ground motion records that have sufficiently large PGAs. by doing so, an extremely small value of μ can also be avoided as far as possible when conducting a parametric study. Therefore, for site classes C and D, only the ground motion records in which the peak value of the x-component, a_{gx0}, is not less than 0.15g were considered. However, less than 50 records with $a_{gx0} \geq 0.05g$ could be found in the PEER-NGA database for site classes B and E; as a result of this limitation, 40 acceleration records were selected for site classes B and E, each with $a_{gx0} \geq 0.05g$, and all of them were non-pulse-like records. Site classes C and D each had 120 acceleration records selected; within those 120 records, there were 90 non-pulse-like records and 30 near-fault pulse-like records in each group. The 90 non-pulse-like records for both site class C and site class D were selected according to different combinations of the magnitude interval and source-to-site distance (defined as the closest distance to the fault rupture zone) interval. Three intervals of the magnitude, M, namely, $5.2 \leq M < 6.0$, $6.0 \leq M < 6.7$ and $6.7 \leq M < 7.7$, and three intervals of the source-to-site distance, D, namely, $0 < D < 14$ km, $14 \leq D < 24$ km and $24 \leq D < 120$ km, were employed herein, resulting in 9 different combinations of magnitude and distance intervals. Given a specific site class (C or D), 10 records were selected for each of these combinations. A total of 320 records of earthquake ground motion, originating from 69 earthquakes with a magnitude M ranging from 5.2 to 7.7, were used in this study. Figure 4.2 shows the distribution of the magnitudes and source-to-site distances of the ground motion records selected for each site class.

The previous section mentioned that the response of the superstructure in the x-direction is influenced by both the y and z (vertical) components of the ground acceleration; therefore, comparing the PGAs of all three components is necessary to

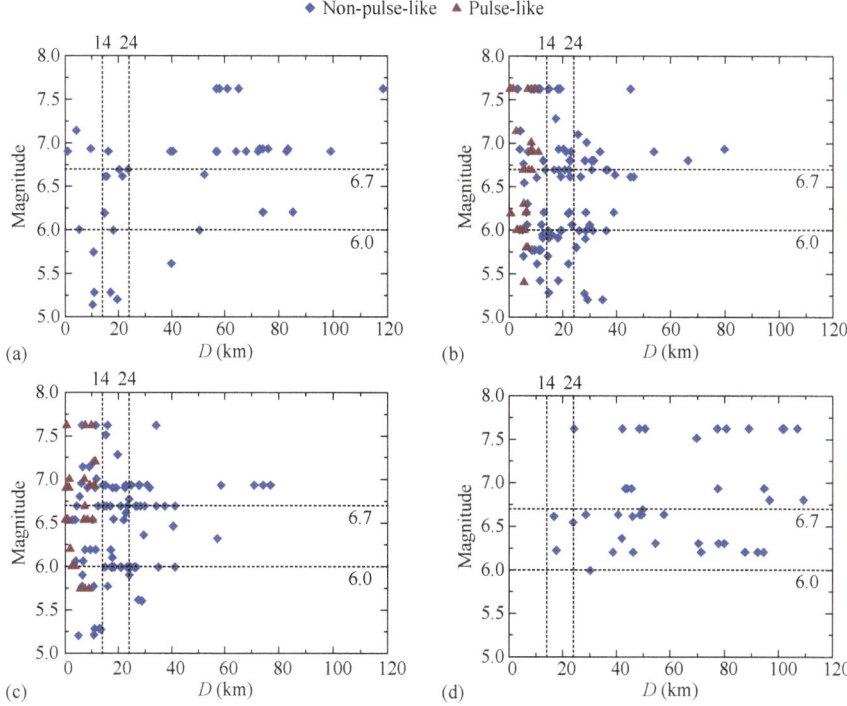

Fig. 4.2 Distribution of magnitudes and source-to-site distances of the ground motion records: **a** site class B; **b** site class C; **c** site class D; and **d** site class E

investigate the significance of the interaction between the two horizontal directions and the impact of the vertical component. Figure 4.3 shows the distributions of a_{gy0}/a_{gx0} and a_{gz0}/a_{gx0} for the 320 selected ground motion records. As shown in Fig. 4.3a, the values of a_{gy0}/a_{gx0} are mostly (95.6%) between 0.5 and 2 with an average value of 1.02. This shows that the assessed earthquake ground motions typically demonstrate comparable intensities in both horizontal directions. Figure 4.3b reveals that the a_{gz0}/a_{gx0} values are primarily concentrated within the 0.2–0.8 range, with an average value of 0.62, which indicates that the PGA of the vertical component is generally smaller than those of the horizontal component.

4.3 Normalized Ground Motion Intensity for the Initiation of Sliding

If the third equation of Eq. (4.1) is satisfied throughout the entire excitation history, sliding will not occur; thus, the critical static friction coefficient, μ_{cr}, for the initiation of sliding can be determined using the following equation:

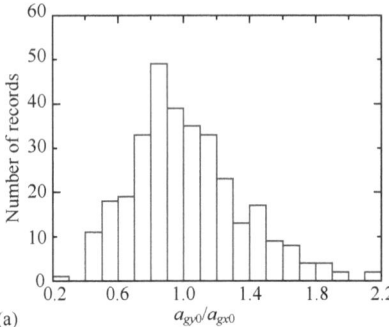 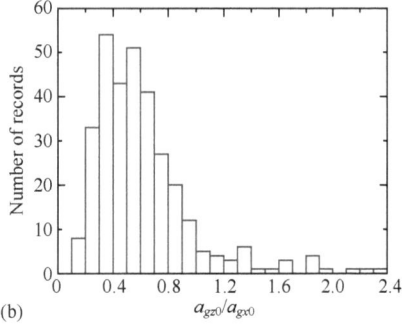

(a)   (b)

Fig. 4.3 Distributions of **a** a_{gy0}/a_{gx0}; and **b** a_{gz0}/a_{gx0}

$$\mu_{cr} = \max_t \left(\frac{\sqrt{\left(\alpha \ddot{u}_{rx}(t) + \ddot{u}_{gx}(t)\right)^2 + \left(\alpha \ddot{u}_{ry}(t) + \ddot{u}_{gy}(t)\right)^2}}{g + \ddot{u}_{gz}(t)} \right) \qquad (4.8)$$

where $\ddot{u}_{rx}(t)$ and $\ddot{u}_{ry}(t)$ are computed using the first and second equations of Eq. (4.1). According to Eq. (4.8), disregarding the influence of the vertical component, $\ddot{u}_{gz}(t)$, there exists a linear correlation between μ_{cr} and PGA; and hence, $\max(a_{gx0}, a_{gy0})/\mu g$ can be used as a more generalized indicator for determining the occurrence of sliding, i.e., sliding occurs when $\max(a_{gx0}, a_{gy0})/\mu_s g > \max(a_{gx0}, a_{gy0})/\mu_{cr} g$.

Figure 4.4 plots the counted median, counted 5th percentile, and counted 95th percentile values of $\max(a_{gx0}, a_{gy0})/\mu_{cr} g$ for the 120 selected ground motion records for site class D. In the computation, T_y is taken to be the same as T_x. Figure 4.4a shows that as the mass ratio, α, decreases, the median value of $\max(a_{gx0}, a_{gy0})/\mu_{cr} g$ increases for a given superstructure period, T_x. The reason for this is that, in short-period structures, the peak relative acceleration is typically larger than the corresponding PGA, leading to larger values of μ_{cr} in Eq. (4.8) for larger values of α. Figure 4.4a also shows that, except for $T_x \leq 0.3$ s, the median value of $\max(a_{gx0}, a_{gy0})/\mu_{cr} g$ increases as T_x increases for a given mass ratio. This result is consistent with the shape of the corresponding response spectrum of FB structures. When $\max(a_{gx0}, a_{gy0})/\mu_s g$ is equal to the corresponding counted 5th percentile value of $\max(a_{gx0}, a_{gy0})/\mu_{cr} g$, sliding occurs for a small number of ground motions. However, the effect of this short-term sliding on the superstructure response is insignificant because of the very short sliding duration in these cases. Therefore, the response of an SB structure can be considered the same as that of the corresponding FB structure when $\max(a_{gx0}, a_{gy0})/\mu_s g$ is smaller than the corresponding value in Fig. 4.4b. During 0.2 s $\leq T_x \leq 0.4$ s, the 5th percentile value of $\max(a_{gx0}, a_{gy0})/\mu_{cr} g$ is estimated to be 0.27 for $\alpha = 0.9$, and it increases to 0.42 when $\alpha = 0.5$. The values in Fig. 4.4c can be regarded as the lower bounds of $\max(a_{gx0}, a_{gy0})/\mu_s g$ to ensure the occurrence of sliding. The trends observed in Fig. 4.4 are also evident in the results

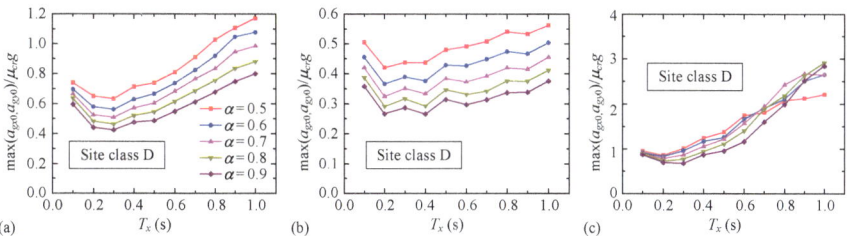

Fig. 4.4 Typical statistical values of $\max(a_{gx0}, a_{gy0})/\mu_{cr}g$: **a** counted median; **b** counted 5th percentile; and **c** counted 95th percentile

for the other site classes, except with slight difference in their respective specific values.

4.4 Parametric Study for the Maximum Superstructure Response

According to the previous discussion on the range of $a_{gx0}/\mu g$, ten levels of $a_{gx0}/\mu g$, namely, 0.25, 0.5, 1, 2, 4, 6, 8, 12, 16 and 20, in which $a_{gx0}/\mu g = 0.25$ is basically equivalent to the FB case, are used in the following analyses. The dynamic friction coefficient, μ, for the sliding interfaces used in SB structures (Barbagallo et al., 2017; Hasani, 1996; Jampole et al., 2016; Nanda et al., 2012; Qamaruddin et al., 1986; Yegian et al., 2004) falls between 0.07 and 0.41. If μ is limited to the range between 0.07 and 0.41, none of the selected ground motion records can yield all the levels of $a_{gx0}/\mu g$ considered. However, in order to analyze the dispersion of the superstructure response at various levels of $a_{gx0}/\mu g$, it is necessary to apply the same number of ground motion records for all $a_{gx0}/\mu g$ levels. For this purpose, the value of μ is adjusted with unscaled ground motion records for each target value of $a_{gx0}/\mu g$. By doing so, to reach a large value of $a_{gx0}/\mu g$ (e.g., $a_{gx0}/\mu g = 16$ or 20) for ground motion records with small PGAs, it will be inevitable to use very small values of μ (e.g., $\mu \leq 0.02$).

Figure 4.5 shows each individual value of $A_x/\mu g$ computed using the 90 non-pulse-like records for site class D in addition to the mean, the mean plus one standard deviation (SD) (corresponding to the 84th percentile value of the normal distribution), the counted median and the counted 84th percentile. In this computation, $\mu_s = \mu$ and $T_x = T_y = 0.3$ s are adopted. The mean and the counted median agree well with each other, as do the mean plus SD and the counted 84th percentile. As stated above, the value of μ is adjusted to reach the target value of $a_{gx0}/\mu g$ with unscaled ground motion records. The considered cases are classified into two groups based on the values of μ obtained; one with μ within the common range of 0.07–0.41 and the other with μ outside this range. Different symbols are used in Fig. 4.5 to denote the data in these two groups. As expected, the resulting values of μ are basically beyond the common

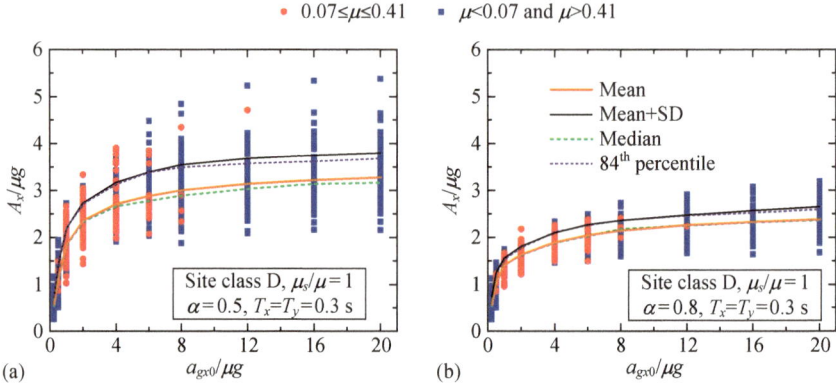

Fig. 4.5 Individual and some statistical values of $A_x/\mu g$: **a** $\alpha = 0.5$; and **b** $\alpha = 0.8$

range when $a_{gx0}/\mu g \geq 12$, because there are very few records of PGA ≥ 0.8g in the ground motion database. In order to determine the reliability of the results obtained using μ which is out of the common range, the probability densities of the computed values of $A_x/\mu g$ at $a_{gx0}/\mu g = 1$ and 4 corresponding to the two different groups are compared in Fig. 4.6; in this figure, the normal probability density functions with the corresponding mean and SD are also presented. The distributions of $A_x/\mu g$ in each group are fundamentally similar to each other. Based on this observation, similar results would likely be obtained for large values of $a_{gx0}/\mu g$ if a sufficient number of ground motion records with large PGAs were used. Additionally, the probability density of the calculated values corresponds quite well with the corresponding fitted normal probability density function, which suggests that the probability distribution of $A_x/\mu g$ for a given $a_{gx0}/\mu g$ value is approximately in accordance with a normal distribution.

4.4.1 Comparison of the Response in Two Orthogonal Directions

Figure 4.7 shows the mean values of $A_x/\mu g$ at each level of $a_{gx0}/\mu g$, in addition to each individual value of $A_y/\mu g$ at the corresponding level of $a_{gy0}/\mu g$ computed using the 90 non-pulse-like records selected for site class D. In this computation, $\mu_s = \mu$ and $T_x = T_y = 0.3$ s are used. Figure 4.7 shows a basically uniform distribution of the discrete points with $(a_{gy0}/\mu g, A_y/\mu g)$ coordinates along both sides of the mean $A_x/\mu g$ versus $a_{gx0}/\mu g$ curve. This indicates that the normalized peak pseudoacceleration and the normalized PGA have an essentially identical relationship in both orthogonal horizontal directions; in other words, the results obtained for the x-direction can also be applied to the y-direction. Therefore, only the response in the x-direction is analyzed hereafter.

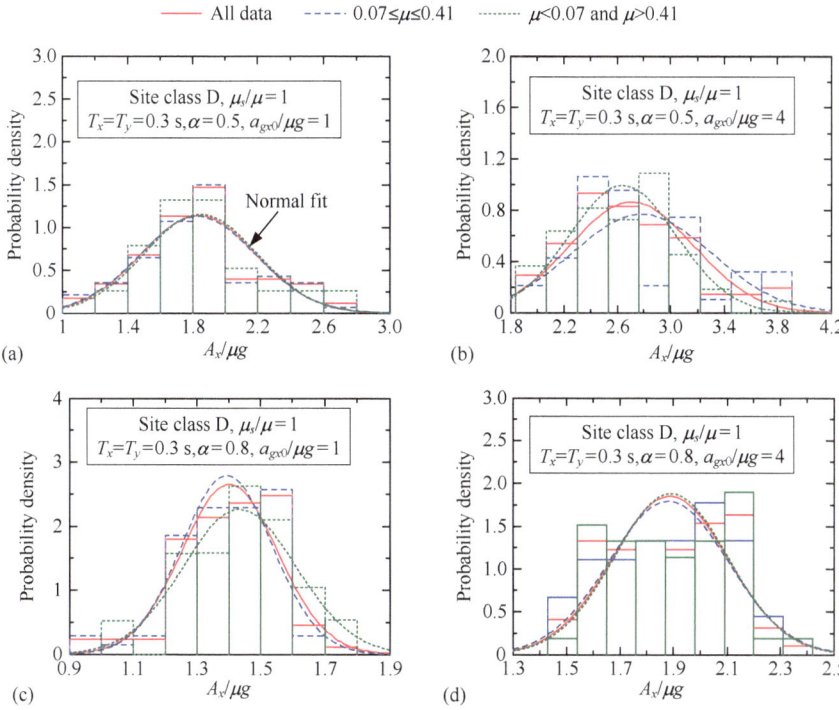

Fig. 4.6 Comparison of probability densities of $A_x/\mu g$ corresponding to different groups: **a** $\alpha = 0.5, a_{gx0}/\mu g = 1$; **b** $\alpha = 0.5, a_{gx0}/\mu g = 4$; **c** $\alpha = 0.8, a_{gx0}/\mu g = 1$; and **d** $\alpha = 0.8, a_{gx0}/\mu g = 4$

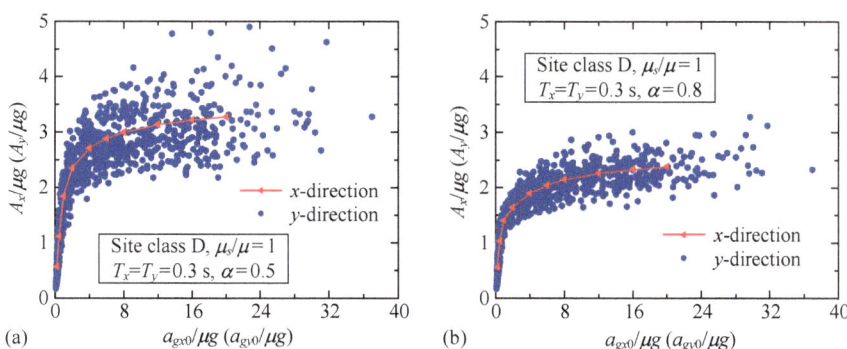

Fig. 4.7 Comparison of the relationship between the normalized peak pseudo-acceleration and the normalized PGA for the two orthogonal horizontal directions: **a** $\alpha = 0.5$; and **b** $\alpha = 0.8$

4.4.2   Effect of the Vertical Earthquake Component

The responses of SB structures under only the two horizontal components of earthquake excitation were also computed in order to study the effect of the vertical component. Figure 4.8 shows the ratios of A_x under three-component excitation to that under the corresponding excitation with the two horizontal components. The 90 non-pulse-like records selected for site class D with $\mu_s = \mu$ and $T_x = T_y$ were used to compute the results shown in Fig. 4.8. According to these figures, the vertical component of ground motion can either increase or decrease the horizontal response of the superstructure. In general, the vertical component has a greater effect on stiffer structures. For certain ground motions, with $T_x \leq 0.3$ s, the ratios of A_x under three-component excitation to that under two-component excitation exceed 1.15. However, for most of the ground motions considered, the ratios of A_x under three-component excitation to that under two-component excitation are between 0.95 and 1.05, and the mean values are basically equal to 1.0. Therefore, the overall effect of the vertical component on the superstructure response is negligible.

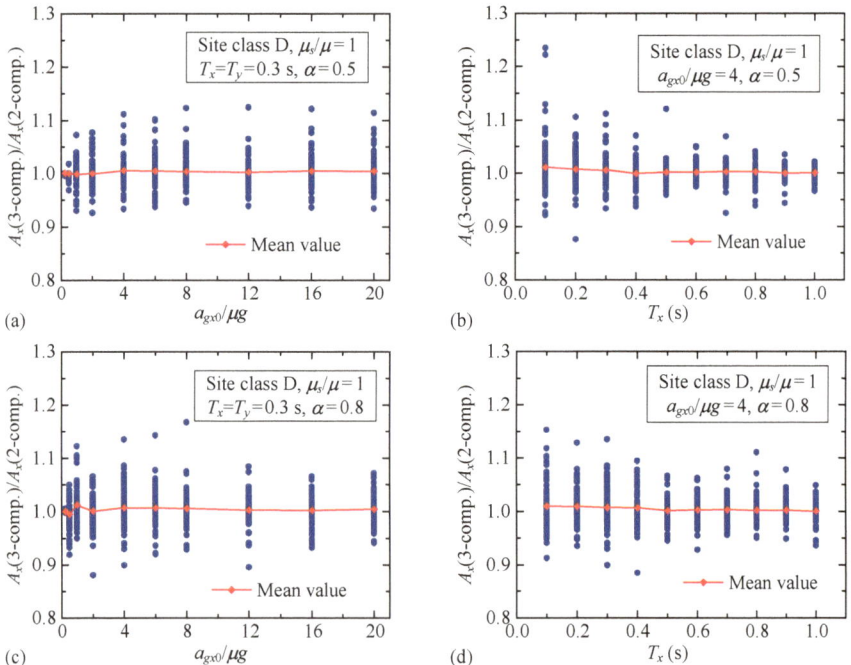

Fig. 4.8 Ratio of A_x under three-component excitation to that under the corresponding two-horizontal-component excitation: **a** $\alpha = 0.5$, $T_x = T_y = 0.3$ s; **b** $\alpha = 0.5$, $a_{gx0}/\mu g = 4$; **c** $\alpha = 0.8$, $T_x = T_y = 0.3$ s; and **d** $\alpha = 0.8$, $a_{gx0}/\mu g = 4$

4.4.3 Effect of the Natural Period of the Superstructure

Figure 4.9 shows the relationship between the mean $A_x/\mu g$ and T_x for different values of α and $a_{gx0}/\mu g$. The mean values of $A_x/\mu g$ were computed using the 90 non-pulse-like records for site class D while assuming that $\mu_s = \mu$ and $T_x = T_y$. According to Fig. 4.9, as T_x increases, the mean values of $A_x/\mu g$ increase for $T_x \leq$ 0.3 s, but decrease for $T_x \geq 0.4$ s, resulting in the maximum mean values of $A_x/\mu g$ are obtained at $T_x = 0.3$ s or 0.4 s. In general, the mean values of $A_x/\mu g$ at $T_x =$ 0.3 s or 0.4 s are close to each other. As $a_{gx0}/\mu g$ increases, the influence of T_x on the superstructure response decreases; for example, the ratio of the minimum to the maximum mean values of $A_x/\mu g$ in Fig. 4.9a is 0.69 for $a_{gx0}/\mu g = 2$ and increases to 0.87 for $a_{gx0}/\mu g = 12$. As shown in Fig. 4.9, it can be inferred that the mean value of $A_x/\mu g$ does not exhibit a considerable variation within the range of T_x that has been considered; therefore, it is appropriate to use the response from the period with the maximum mean $A_x/\mu g$ as a representation of the responses for possible SB structures. As mentioned above, for site class D, this period can be taken as 0.3 s; for site classes B, C, and E, the critical periods obtained are 0.2, 0.2, and 0.4 s, respectively.

In all the above analyses, $T_x = T_y$ is adopted. Figure 4.10 compares the mean values of $A_x/\mu g$ for different values of T_x/T_y in order to investigate the possible impact of T_x/T_y on the superstructure response. In general, the mean values of $A_x/\mu g$ at a given level of $a_{gx0}/\mu g$ decreases as T_x/T_y increases. Nevertheless, this variation is quite limited; the presented results in Fig. 4.10 show that the ratios of the mean value of $A_x/\mu g$ for $T_x/T_y = 0.5$ to $T_x/T_y = 2$ do not exceed 1.06. Therefore, the results obtained for $T_x = T_y$ are representative of those obtained for the possible value of T_x/T_y in the range considered.

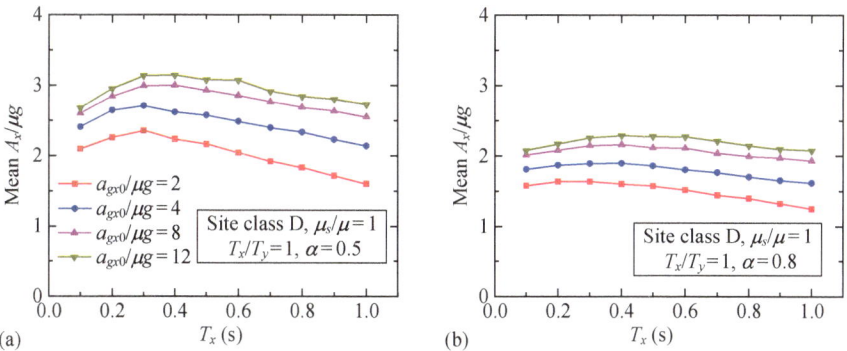

(a) (b)

Fig. 4.9 Relationship between mean $A_x/\mu g$ and T_x: **a** $\alpha = 0.5$; and **b** $\alpha = 0.8$

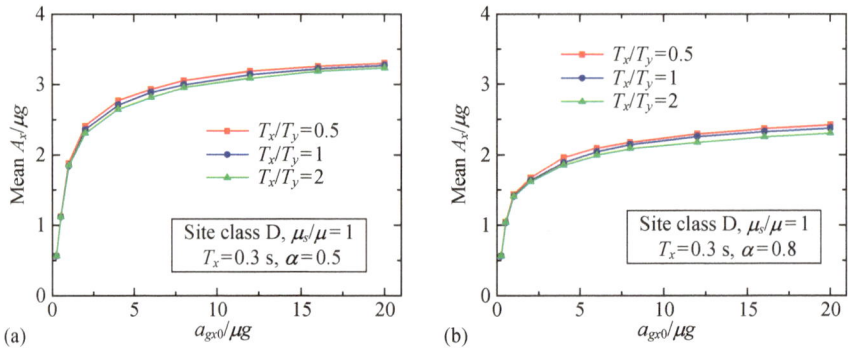

(a)   (b)

Fig. 4.10 Effect of T_x/T_y on the mean values of $A_x/\mu g$: **a** $\alpha = 0.5$; and **b** $\alpha = 0.8$

4.4.4 Effect of the Difference Between the Static and Dynamic Friction Coefficients

Based on the investigation mentioned above, μ_s/μ generally within the range of 1.0–1.4. Figure 4.11 compares the mean values of $A_x/\mu g$ corresponding to different values of μ_s/μ in order to investigate the influence of μ_s/μ. As expected, for most of the ground motions considered, the superstructure response is not significantly influenced by the value of μ_s/μ when $a_{gx0}/\mu g = 0.25$, because sliding does not occur at this $a_{gx0}/\mu g$ level. The influence of μ_s/μ is most clearly observed for $a_{gx0}/\mu g = 1$. If $a_{gx0}/\mu g$ exceeds 1, the influence of μ_s/μ decreases as $a_{gx0}/\mu g$ increases because for larger $a_{gx0}/\mu g$ values, the responses of SB structures are primarily dominated by the sliding phase, during which the responses are independent of the static friction coefficient. The ratios of the mean value of $A_x/\mu g$ for $\mu_s/\mu = 1.4$ to that for $\mu_s/\mu = 1$ are all below 1.08 and mostly below 1.02. Therefore, the effect of the difference between the static and dynamic friction coefficients can be neglected. In the following analyses, $\mu_s = \mu$ is assumed.

4.4.5 Effects of the Earthquake Magnitude and Source-to-Site Distance

The mean values of $A_x/\mu g$ for each distance interval and magnitude were computed using 90 non-pulse-like records selected for site class D in order to study the effects of the earthquake magnitude and source-to-site distance. Figure 4.12 shows the mean values of $A_x/\mu g$ for the three magnitude intervals. For the entire range of $a_{gx0}/\mu g$, no general trend can be observed for $A_x/\mu g$ as the earthquake magnitude increases. However, when $a_{gx0}/\mu g = 0.25$, the mean values of $A_x/\mu g$ for $6.7 \le M < 7.7$ are approximately 10% larger than those for $6.0 \le M < 6.7$; the relative differences in

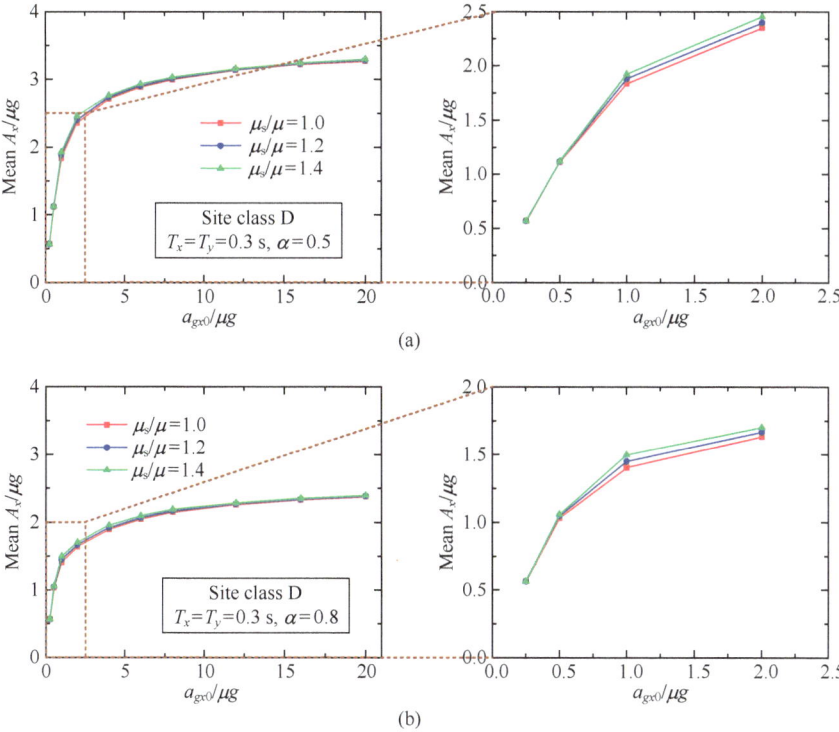

Fig. 4.11 Effect of μ_s/μ on the mean values of $A_x/\mu g$: **a** $\alpha = 0.5$; and **b** $\alpha = 0.8$

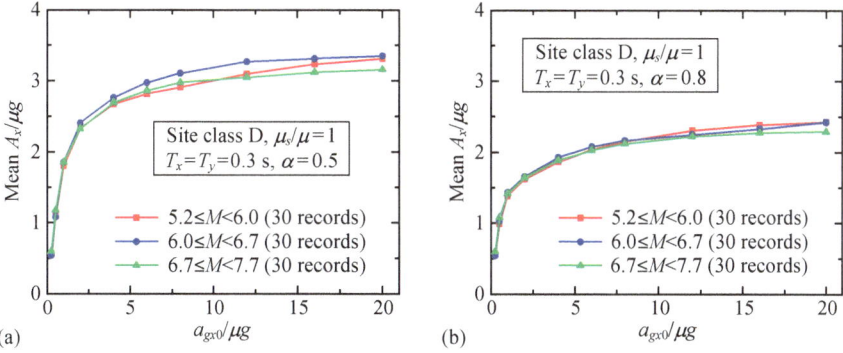

Fig. 4.12 Mean values of $A_x/\mu g$ for three magnitude intervals: **a** $\alpha = 0.5$; and **b** $\alpha = 0.8$

the mean values of $A_x/\mu g$ between any two of these groups at a given $a_{gx0}/\mu g$ are all below 7% and mostly below 5%. This indicates that the earthquake magnitude has little effect on the superstructure responses of SB structures.

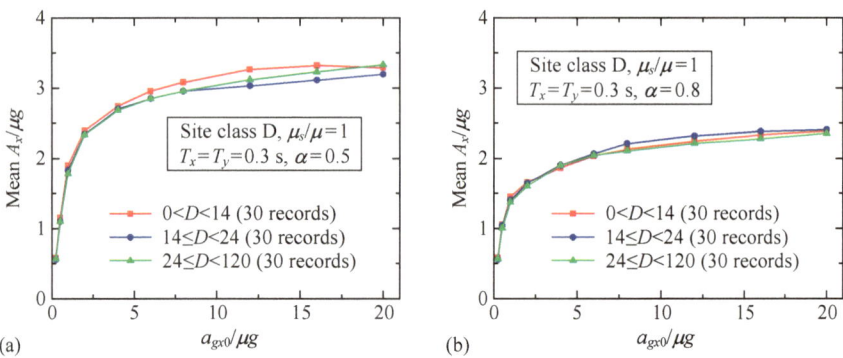

Fig. 4.13 Mean values of $A_x/\mu g$ for three distance intervals: **a** $\alpha = 0.5$; and **b** $\alpha = 0.8$

Figure 4.13 shows the mean values of $A_x/\mu g$ for three distance intervals. The influence of the source-to-site distance on the superstructure response, like that of the magnitude of the earthquake, is also insignificant.

4.4.6 Effect of Near-Fault Pulses

The acceleration, velocity, and displacement histories of near-fault ground motions influenced by forward directivity contain distinct pulses. To investigate the possible effects of these pulses, the mean values of $A_x/\mu g$ computed using the 30 pulse-like records and 30 non-pulse-like records both recorded for $0 < D < 14$ km and site class D are compared in Fig. 4.14. When sliding basically does not occur, i.e., $a_{gx0}/\mu g = 0.25$, the mean value of $A_x/\mu g$ corresponding to the non-pulse-like records is 1.16 times that corresponding to the pulse-like records. The corresponding FB structure has a larger response amplification factor, A_x/a_{gx0}, for non-pulse-like ground motions compared to pulse-like ones, as indicated. Chopra and Chintanapakdee (2001) reported similar results, where they examined the normalized response spectra of harmonic excitations containing different numbers of cycles to interpret this phenomenon. The response amplification factor increased as the number of cycles increased, implying that the response amplification factors of pulse-like ground motions with one or several dominant pulses are generally smaller than those of non-pulse-like ground motions with more excitation cycles. As illustrated in Fig. 4.14a, when $\alpha = 0.5$, the ratio of the mean value of $A_x/\mu g$ for the non-pulse-like records to that of the pulse-like records remains almost the same as $a_{gx0}/\mu g$ increases. However, when α increases to 0.8, the ratio decreases significantly as $a_{gx0}/\mu g$ increases, as shown in Fig. 4.14b. This phenomenon is consistent with the fact that the difference in the superstructure response between different ground motions decreases as α increases. More detailed results related to this fact are presented in the next section.

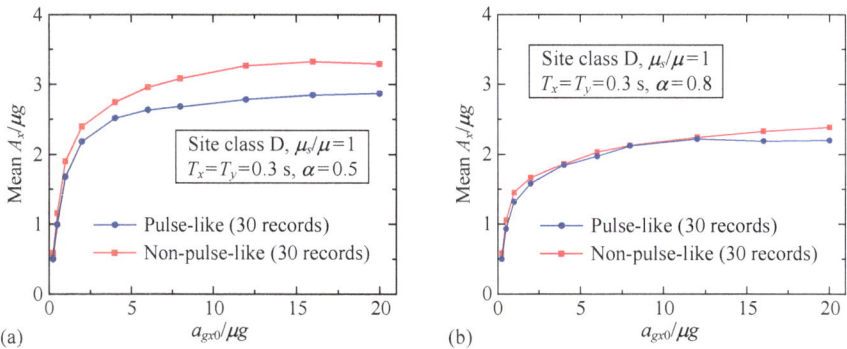

Fig. 4.14 Effect of near-fault pulses on the mean values of $A_x/\mu g$: **a** $\alpha = 0.5$; and **b** $\alpha = 0.8$

4.4.7 Statistical Results for Each Site Class

Figure 4.15 shows the mean values of $A_x/\mu g$ for each site class. Herein, only the non-pulse-like records were used for the computation to maintain consistency. Thus, the numbers of ground motion records used for site classes B, C, D and E are 40, 90, 90 and 40, respectively. As shown in Fig. 4.15, except for $a_{gx0}/\mu g = 0.25$ (which is basically equivalent to the FB case), as the mass ratio, α, increases, the superstructure response reduces. This phenomenon can be easily explained by using the governing equations for the sliding phases during unidirectional excitation. Under unidirectional excitation in the x-direction, Eq. (4.2) can be simplified into

$$\ddot{u}_{rx} + 2\frac{\xi_x}{\sqrt{1-\alpha}}\frac{\omega_x}{\sqrt{1-\alpha}}\dot{u}_{rx} + \frac{\omega_x^2}{1-\alpha}u_{rx} = \frac{\dot{u}_{sx}}{|\dot{u}_{sx}|}\frac{\mu g}{1-\alpha} \qquad (4.9)$$

which is the differential equation of a single-degree-of-freedom (SDOF) system with a natural frequency of $\omega_x/\sqrt{1-\alpha}$ and a damping ratio of $\xi_x/\sqrt{1-\alpha}$ subjected to a step force corresponding to a static displacement of $\mu g/\omega_x^2$. The damping ratio of this equivalent system increases as α increases, leading to a general decrease in the response of u_{rx} as α increases.

After $a_{gx0}/\mu g$ exceeds 0.5, the mean $A_x/\mu g$ versus $a_{gx0}/\mu g$ curves experience a rapid decline in tangent slopes as a result of sliding. When $a_{gx0}/\mu g$ exceeds a sufficiently large value, i.e., there is an upper limit for the superstructure response of an SB structure, the tangent slopes are expected to finally become 0. This situation is favorable for isolating extremely large earthquakes. The efficiency of the SB system can also be demonstrated by the value of A_x/a_{gx0}, which is equal to the origin-oriented secant slope of the $A_x/\mu g$ versus $a_{gx0}/\mu g$ curve. Taking $\alpha = 0.8$ in Fig. 4.15b (site class C) as an example: when $a_{gx0}/\mu g = 2$, the mean value of A_x/a_{gx0} is 0.82, whereas this value is 2.23 for the FB case; consequently, the superstructure response of the SB structure is just 36.8% of that of the corresponding FB structure in this instance.

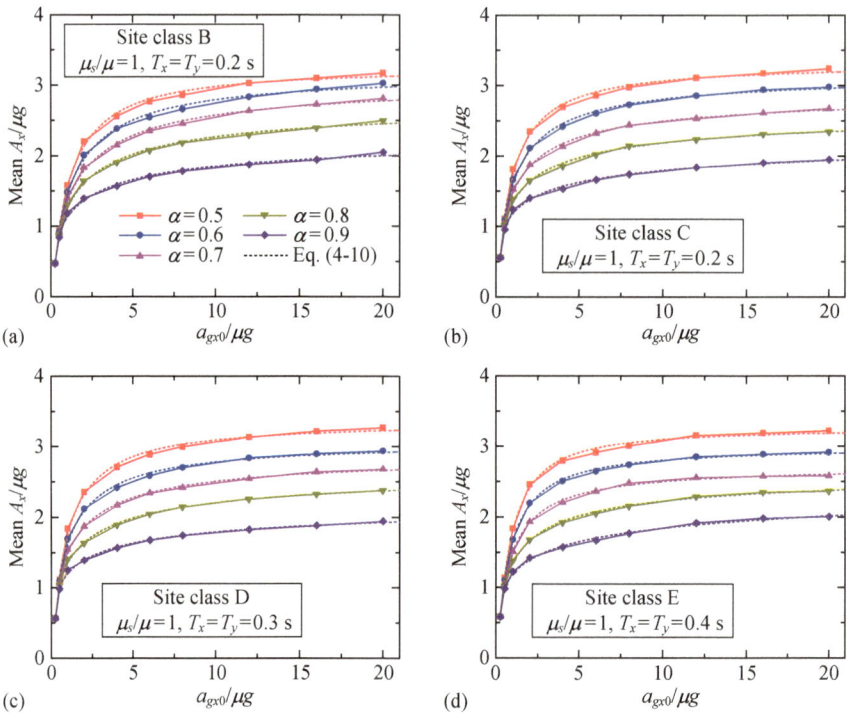

Fig. 4.15 Mean values of $A_x/\mu g$ for each site class: **a** site class B; **b** site class C; **c** site class D; and **d** site class E

Figure 4.16 presents the ratios of the mean value of $A_x/\mu g$ for site class C (D or E) to that for site class B, for $\alpha = 0.5$ and 0.8, in order to investigate the effects of local site conditions on the superstructure response. It is evident that the response of the superstructure is affected by the local site conditions; as the site soil becomes softer, the mean values of $A_x/\mu g$ increase. For $a_{gx0}/\mu g = 0.25$, when sliding basically does not occur, the ratios of the mean values of $A_x/\mu g$ for site classes C, D and E to that for site class B are equal to 1.19, 1.20 and 1.24, respectively. These ratios generally decrease as $a_{gx0}/\mu g$ increases. As shown in Fig. 4.16a, when $a_{gx0}/\mu g = 2$, these ratios decrease to 1.07, 1.07 and 1.12 for site classes C, D and E, respectively; and they further decrease to 1.02, 1.03 and 1.04 when $a_{gx0}/\mu g = 12$.

Figure 4.17 shows the coefficients of variation (COVs) of $A_x/\mu g$ for every site class in order to investigate the dispersion of the superstructure response at a specified value of $a_{gx0}/\mu g$. The COVs of the $A_x/\mu g$ versus $a_{gx0}/\mu g$ curves are similar for all four site classes. In the range of $a_{gx0}/\mu g \leq 2$, the COVs of $A_x/\mu g$ decrease rapidly as $a_{gx0}/\mu g$ increases; after $a_{gx0}/\mu g$ exceeds 2, the COVs of $A_x/\mu g$ are quite constant. This means that sliding tends to reduce the dispersion of the superstructure response

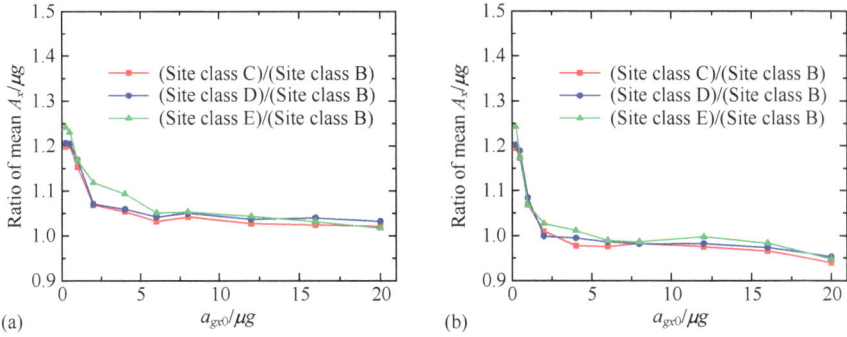

Fig. 4.16 Effect of local site conditions on the mean values of $A_x/\mu g$: **a** $\alpha = 0.5$; and **b** $\alpha = 0.8$

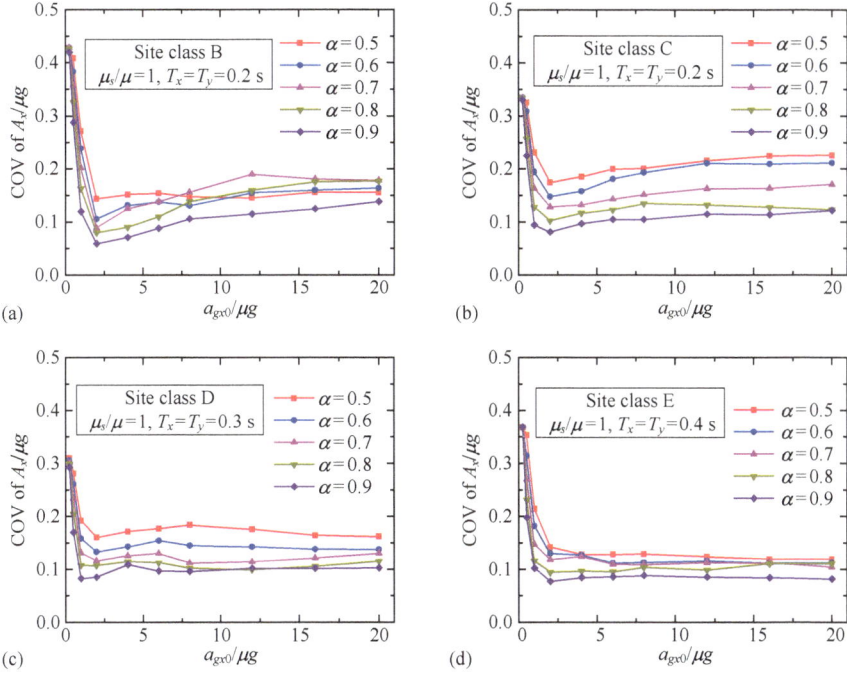

Fig. 4.17 Coefficients of variation of $A_x/\mu g$ for each site class: **a** site class B; **b** site class C; **c** site class D; and **d** site class E

due to the record-to-record variability. As α increases, the COVs of $A_x/\mu g$ decrease in general. This can also be interpreted by using Eq. (4.9); larger values of α leads to larger equivalent damping ratios for the sliding phases, further resulting in smaller dispersion of the structural response.

4.5 Simplified Equations for Estimating the Maximum Superstructure Response

For the design of SB structures, it is desirable to employ simplified equations to estimate the peak pseudoacceleration of the superstructure. Based on the preceding discussions, we know the following: (1) the relationship between $A_x/\mu g$ and $a_{gx0}/\mu g$ is basically identical to that between $A_y/\mu g$ and $a_{gy0}/\mu g$; (2) the dependencies of the mean value of $A_x/\mu g$ (or $A_y/\mu g$) on the vertical earthquake component, T_x/T_y, μ_s/μ, the earthquake magnitude and the source-to-site distance can be neglected; and (3) the response of possible SB structures can be represented by the response at a critical period for each site class with appropriate conservativeness. Thus, the following equation was proposed to estimate the mean values of $A_x/\mu g$ and $A_y/\mu g$:

$$\frac{A_x}{\mu g} = \frac{\beta_1 \left(a_{gx0}/\mu g\right)^{\beta_2}}{\left(a_{gx0}/\mu g\right)^{\beta_2} + \beta_3} \quad \text{and} \quad \frac{A_y}{\mu g} = \frac{\beta_1 \left(a_{gy0}/\mu g\right)^{\beta_2}}{\left(a_{gy0}/\mu g\right)^{\beta_2} + \beta_3} \tag{4.10}$$

where β_1, β_2 and β_3 are the regression coefficients that depend on the site class and the mass ratio, α. Equation (4.10) captures the trend of $A_x/\mu g$ $\left(A_y/\mu g\right)$ with respect to $a_{gx0}/\mu g$ $\left(a_{gy0}/\mu g\right)$, i.e., $A_x/\mu g \rightarrow 0$ when $a_{gx0}/\mu g \rightarrow 0$, and $A_x/\mu g$ approaches an upper limit when $a_{gx0}/\mu g \rightarrow +\infty$. The Curve Fitting Toolbox of MATLAB (2014) was used to conduct nonlinear regression analyses for determining the values of the regression coefficients in Eq. (4.10). In order to assess conservativeness, regression analyses were conducted using results obtained from non-pulse-like records, considering that responses under pulse-like ground motions generally exhibit smaller values compared to those under non-pulse-like ground motions. Table 4.1 presents the values of β_1, β_2 and β_3 obtained for each site class and various values of α. The values predicted through Eq. (4.10) are also compared to the values computed from response history analyses shown in Fig. 4.15. The mean values of $A_x/\mu g$ can be accurately estimated by the proposed equation. The values of β_1, β_2 and β_3 for the α values not included in Table 4.1 can be found by linear interpolation of the values from Table 4.1, because each regression coefficient has an approximately linear relationship with α; if $\alpha > 0.9$, using the β_1, β_2 and β_3 values for $\alpha = 0.9$ can give conservative results.

As shown in the previous section, a normal distribution is appropriate for modeling the probability distribution of $A_x/\mu g$ with respect to a specific site class, $a_{gx0}/\mu g$ and α. Therefore, if we can further derive a simplified equation for the COV (or SD = mean × COV), then the value of $A_x/\mu g$ (or $A_y/\mu g$) corresponding to any probability of nonexceedance can be readily determined. Figure 4.17 shows that, despite possible variations of the exact values among different groups, the trends of the COVs of $A_x/\mu g$ with respect to $a_{gx0}/\mu g$ remain consistent for the four site classes. Because the computed COV values are related to the selected ground motion records used in the computation (i.e., a different set of records for the same site class may lead to different COVs), it is reasonable to expect equivalent dispersion levels for the

Table 4.1 Values of the regression coefficients in Eq. (4.10)

Site class	α	β_1	β_2	β_3
B	0.5	3.26	1.07	1.12
	0.6	3.20	0.92	1.22
	0.7	3.10	0.81	1.30
	0.8	2.80	0.73	1.24
	0.9	2.41	0.58	1.15
C	0.5	3.32	1.03	0.90
	0.6	3.17	0.88	0.97
	0.7	2.92	0.76	0.97
	0.8	2.71	0.63	1.02
	0.9	2.48	0.45	1.10
D	0.5	3.37	1.01	0.92
	0.6	3.09	0.90	0.90
	0.7	2.93	0.76	0.97
	0.8	2.78	0.61	1.06
	0.9	2.39	0.47	0.99
E	0.5	3.27	1.16	0.81
	0.6	3.00	1.06	0.80
	0.7	2.73	0.94	0.81
	0.8	2.68	0.68	0.99
	0.9	2.87	0.40	1.41

four site classes, provided that a sufficient number of records are selected for each group. In general, the computed COV values are the largest for site class C under the circumstance of the ground motion records considered. Therefore, the data of site class C are used to derive the simplified equation for the COVs of $A_x/\mu g$ and $A_y/\mu g$ since a larger COV value leads to a conservative result for a probability of nonexceedance larger than 50%. The proposed equation is given by

$$\delta_{A_x/\mu g} = \gamma_1 \exp\left(-\gamma_2\left(a_{gx0}/\mu g\right)\right) + \gamma_3 \quad \text{and} \quad \delta_{A_y/\mu g} = \gamma_1 \exp\left(-\gamma_2\left(a_{gy0}/\mu g\right)\right) + \gamma_3$$
$$(4.11)$$

where $\delta_{A_x/\mu g}$ $\left(\delta_{A_y/\mu g}\right)$ is the COV of $A_x/\mu g$ $\left(A_y/\mu g\right)$, and γ_1, γ_2 and γ_3 are the regression coefficients that depend on α. Table 4.2 presents the values of γ_1, γ_2 and γ_3 based on the nonlinear regression analyses for various α values. Figure 4.18 shows a comparison of the values predicted using Eq. (4.11) and the values obtained through response history analyses. A satisfactory estimation for the COVs of $A_x/\mu g$ is yielded by the proposed equation. Similar to Eq. (4.10), for the values of α not listed in Table 4.2, the values of γ_1, γ_2 and γ_3 can be calculated from the linear interpolation

Table 4.2 Values of the regression coefficients in Eq. (4.11)

α	γ_1	γ_2	γ_3
0.5	0.23	1.82	0.20
0.6	0.28	2.25	0.19
0.7	0.34	2.25	0.15
0.8	0.42	2.62	0.12
0.9	0.52	3.23	0.10

Fig. 4.18 Comparison between the COVs of $A_x/\mu g$ computed using Eq. (4.11) and those computed from response history analyses

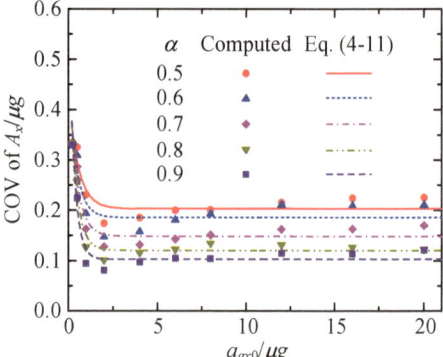

of those provided in Table 4.2; for $\alpha > 0.9$, the values of γ_1, γ_2 and γ_3 for $\alpha = 0.9$ can be used.

4.6 Conclusions

In this chapter, a comprehensive parametric investigation of the normalized peak pseudoacceleration of single-story SB structures subjected to three-component earthquake excitation is presented. The relationship between the normalized peak pseudoacceleration and the normalized PGA is basically identical for the two orthogonal horizontal directions. The horizontal response of the superstructure can be either reduced or increased by the vertical component of ground motion. If $T_x \leq 0.3$ s, for certain ground motions, the superstructure response can increase by more than 1.15 due to the vertical component; but the effect of the vertical component is negligible for the majority of situations. The normalized peak pseudoacceleration exhibits a pattern of initially increasing and subsequently decreasing as the natural period of the superstructure increases. For the range of T_x considered, the variation in the mean value of $A_x/\mu g$ is not very significant. For simplicity and conservativeness, the response of possible SB structures can be represented by the response at the period where the maximum mean $A_x/\mu g$ is generally obtained. The influence of the natural period ratio in the two orthogonal horizontal directions and the possible difference between the static and dynamic friction coefficients on the superstructure responses of SB structures is insignificant.

The effects of the earthquake magnitude and the source-to-site distance are very small and can be neglected in practice. Superstructures typically exhibit smaller responses when subjected to pulse-like ground motions compared to non-pulse-like ones. Local site conditions have an effect on the response of the superstructure. For sites located on softer soil, a larger response is obtained, and the dependence on the local site conditions decreases as the normalized PGA increases. The trend of the COVs of $A_x/\mu g$ with respect to $a_{gx0}/\mu g$ is similar among the four site classes. The COVs decline rapidly at smaller values of $a_{gx0}/\mu g$ and remain basically constant after $a_{gx0}/\mu g \geq 2$. For a given site class, $a_{gx0}/\mu g$ and α, a normal distribution is appropriate for modeling the probability distribution of $A_x/\mu g$.

The mean values and COVs of $A_x/\mu g$ decrease as α increases. An upper limit for the superstructure response exists for every mass ratio, which is beneficial for the isolation of extremely large earthquakes. Implementing Eqs. (4.10) and (4.11) with the associated values of the regression coefficients can provide good estimates for the mean values and COVs, respectively, of $A_x/\mu g$ and can be used to predict the value of the normalized peak pseudoacceleration corresponding to any probability of nonexceedance.

Chapter 5
Equivalent Lateral Forces for Design of Multistory Sliding Base Structures

5.1 Model Descriptions

In Chap. 4, the parameter mass ratio, α, was introduced. This parameter is defined as the ratio between the superstructure mass and the overall mass of the SB structure. The response of the superstructure is heavily influenced by the parameter α, as demonstrated in Chap. 4. The mass ratio of a multistory SB structure presented in Fig. 5.1 can be calculated by

$$\alpha = \frac{\sum_{i=1}^{N} m_i}{\sum_{i=1}^{N} m_i + m_b} \tag{5.1}$$

in which m_i is the mass of the ith floor; m_b represents the mass of the sliding base; and N corresponds to the story number.

Based on practical applications, a maximum of five stories can be considered as the upper limit for the story number of SB structures. For a building with $N \leq 5$, it is reasonable to consider that the mass of every floor is equal. Thus, Eq. (5.1) becomes

$$\alpha = \frac{Nm}{Nm + m_b} \tag{5.2}$$

in which m represents the mass of each floor. In practical applications, the value of α should be at least 0.5 as m is usually equal to or greater than m_b. Hence, the value of α can be considered not less than 0.5.

It is assumed that the stiffness of each story is equal. The parametric study in Chap. 4 demonstrates that the natural period ratio between the two orthogonal horizontal directions has a negligible impact on the superstructure responses. Therefore, it is assumed that the story stiffness in both the x and y directions is identical

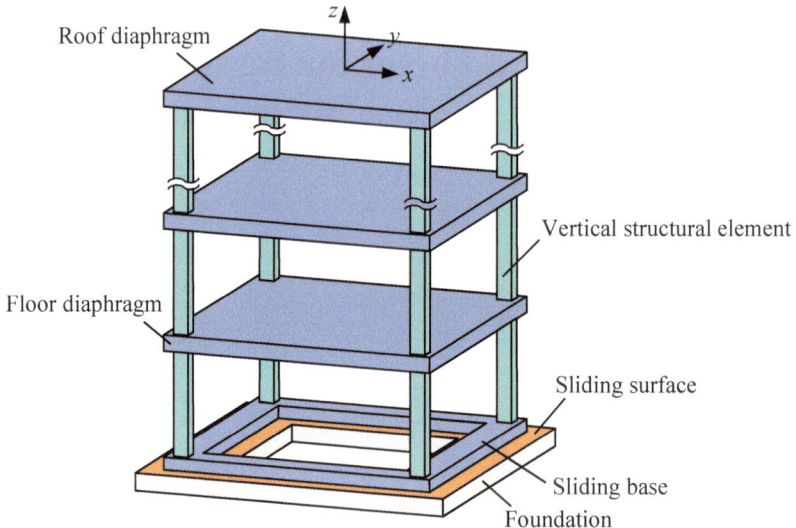

Fig. 5.1 Schematic plot of a multistory SB structure

and represented as k. Given these assumptions, the fundamental period, T_1, of the superstructure is expressed as follows:

$$T_1 = 2\pi \sqrt{\frac{m}{Ck}} \qquad (5.3)$$

in which the values of coefficient C are equal to 0.382, 0.198, 0.121, and 0.081 for N = 2, 3, 4, and 5, respectively. As Chap. 4 demonstrates, the superstructure responses in single-story SB structures remain basically unchanged when the superstructure period falls within the typical range, and it is feasible to conservatively represent the responses of potential SB structures with the response at the period that generally yields the highest superstructure responses. The truth of this result persists in multistory SB structures. The computed results based on $T_1 = 0.3$ s are used in the subsequent analyses as maximum superstructure responses occur mostly at T_1 close to 0.3 s.

The construction of the damping matrix involves the utilization of Rayleigh damping. For $N = 2$, the damping ratios for the first and second modes are taken as 5%, while for $N = 3$ and 4, the damping ratios for the first and third modes are 5%, and for $N = 5$, the damping ratios for the first and fourth modes are taken as 5%.

The static and dynamic friction coefficients are assumed to be the same as the difference between them was found to have little effect on the superstructure responses. In accord with Chap. 4, the maximum value of $a_{gx0}/\mu g$ is taken as 20, where a_{gx0} is the peak value of the x component of the ground acceleration, μ is the friction coefficient, and g is the gravity acceleration.

It was found that the variance between the static and dynamic friction coefficients had minimal impact on the superstructure responses, thus they are assumed to be equal. In accordance with Chap. 4, an upper limit of 20 is placed on the value of $a_{gx0}/\mu g$.

5.2 Peak Base Shear

After obtaining the displacement history by performing response history analysis, the equivalent static forces in the x-direction, $\mathbf{F}_x = [F_{x1}, F_{x2}, \ldots, F_{xN}]$ (F_{xi} is the force acting on the ith floor) for a multistory SB structure can be determined using the following equation (Chopra, 2001)

$$\mathbf{F}_x(t) = \mathbf{k}_x \mathbf{u}_{rx}(t) \tag{5.4}$$

The peak base shear, V_{bx}, can subsequently be computed by

$$V_{bx} = \max_t \left| \mathbf{1}^T \mathbf{k}_x \mathbf{u}_{rx}(t) \right| \tag{5.5}$$

Because of the close relationship between the peak base shear and the mass of the superstructure, as well as the friction coefficient, the normalized peak base shear, $\overline{V}_{bx}$, is introduced as follows

$$\overline{V}_{bx} = \frac{V_{bx}}{Nmg\mu} = \frac{\max_t \left| \mathbf{1}^T \mathbf{k}_x \mathbf{u}_{rx}(t) \right|}{Nmg\mu} \tag{5.6}$$

For a single-story SB structure, Eq. (5.6) is simplified to

$$\overline{V}_{bx} = \frac{\max_t |k u_{rx}(t)|}{mg\mu} = \frac{\omega_x^2 \times \max(|u_{rx}(t)|)}{\mu g} = \frac{A_x}{\mu g} \tag{5.7}$$

As indicated by Eq. (5.7), the normalized peak base shear, $\overline{V}_{bx}$ is equivalent to the normalized peak pseudoacceleration, $A_x/\mu g$, for single-story SB structures. The normalized peak pseudoacceleration has been thoroughly examined in Chap. 4. Hence, with certain modifications, the equations created for predicting $A_x/\mu g$ could be employed to predict $\overline{V}_{bx}$ for multistory SB structures.

By replacing $A_x/\mu g$ in Eqs. (4.10) and (4.11) with $\overline{V}_{bx}$, the following equations are yielded:

$$\overline{V}_{bx} = \frac{\beta_1 \left(a_{gx0}/\mu g \right)^{\beta_2}}{\left(a_{gx0}/\mu g \right)^{\beta_2} + \beta_3} \tag{5.8}$$

and

$$\delta_{\overline{V}_{bx}} = \gamma_1 \exp\left(-\gamma_2\left(a_{gx0}/\mu g\right)\right) + \gamma_3 \tag{5.9}$$

in which $\delta_{\overline{V}_{bx}}$ represents the coefficient of variation (COVs) of $\overline{V}_{bx}$. In order to verify if Eqs. (5.8) and (5.9) are applicable to multistory SB structures, the mean values and COVs of $\overline{V}_{bx}$ of structures with varying story number N are compared in Fig. 5.2a, b, respectively, using the 180 non-pulse-like ground motion records for site classes C and D. As can be seen in Fig. 5.2a, when $a_{gx0}/\mu g$ is given, the mean value of $\overline{V}_{bx}$ decreases and ultimately approaches a constant value as N increases. In order to account for this impact, the introduction of a reduction factor, denoted as γ_N, is necessary. This factor is calculated by taking the ratio of the mean $\overline{V}_{bx}$ of a structure that has N stories, to the mean $\overline{V}_{bx}$ of a single-story structure that has the same α under the same $a_{gx0}/\mu g$. The analyzed data suggest that the influence of α and $a_{gx0}/\mu g$ on γ_N is minimal. The nonlinear regression analyses yielded the following formula for calculating γ_N:

$$\gamma_N = 0.25e^{-0.65N} + 0.86 \tag{5.10}$$

Therefore, the mean value of $\overline{V}_{bx}$ for an N-story SB structure can be calculated by the following equation:

$$\overline{V}_{bx} = \left(0.25e^{-0.65N} + 0.86\right)\frac{\beta_1\left(a_{gx0}/\mu g\right)^{\beta_2}}{\left(a_{gx0}/\mu g\right)^{\beta_2} + \beta_3} \tag{5.11}$$

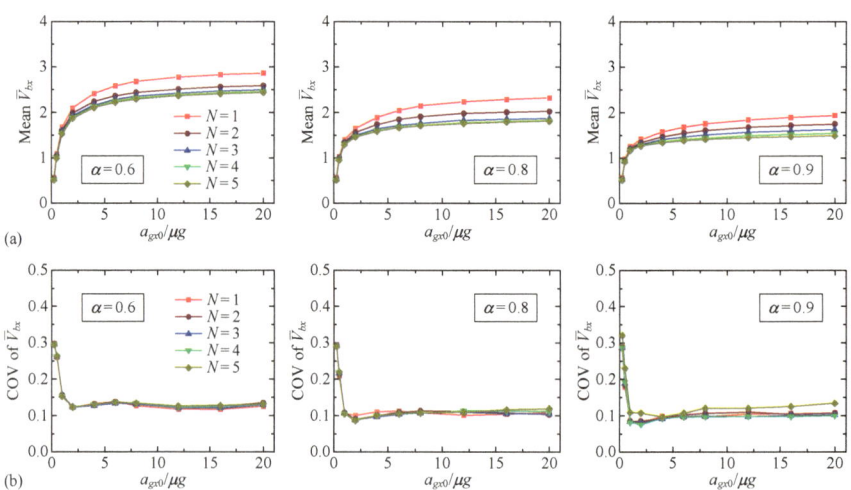

Fig. 5.2 Statistical values of $\overline{V}_{bx}$: **a** mean values; **b** COVs

The values of the coefficients β_1, β_2 and β_3 are provided in Table 4.1. As depicted in Fig. 5.2b, The COVs of $\overline{V}_{bx}$ are not significantly affected by the story number N. Hence, Eq. (5.9) can be used for multistory SB structures without requirement of modifications. The values of γ_1, γ_2 and γ_3 are provided in Table 4.2.

5.3 Equivalent Lateral Force Distribution

The response histories of $V_{xi}/(Nmg)$ for a three-story SB structure (with $\alpha = 0.8$ and $\mu = 0.1$) and corresponding FB structure under the Mammoth Lakes record ($a_{gx0} = 0.39g$) from the 1980 Mammoth Lakes earthquake are presented in Fig. 5.3, where V_{xi} is the story shear of the ith story. In the case of the FB structure, the peak story shears for various stories occur simultaneously. However, for the SB structure, the peak shear time for each story varies. Hence, it is not feasible to employ the distribution of equivalent static forces at the peak base shear to ascertain the peak shear of other stories.

Peak story shears are the response quantities that need to be used in design. There-fore, the equivalent lateral forces were calculated through the following process: (1)

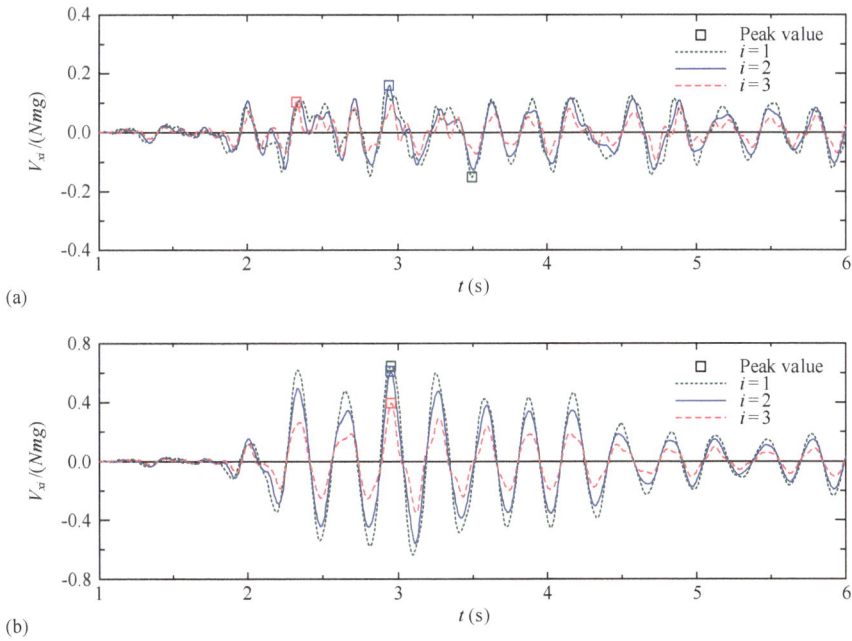

Fig. 5.3 Response histories of each story shear of **a** three-story SB structure (with $\alpha = 0.8$ and $\mu = 0.1$) and **b** three-story FB structure subjected to the Mammoth Lakes record from the 1980 Mammoth Lakes earthquake

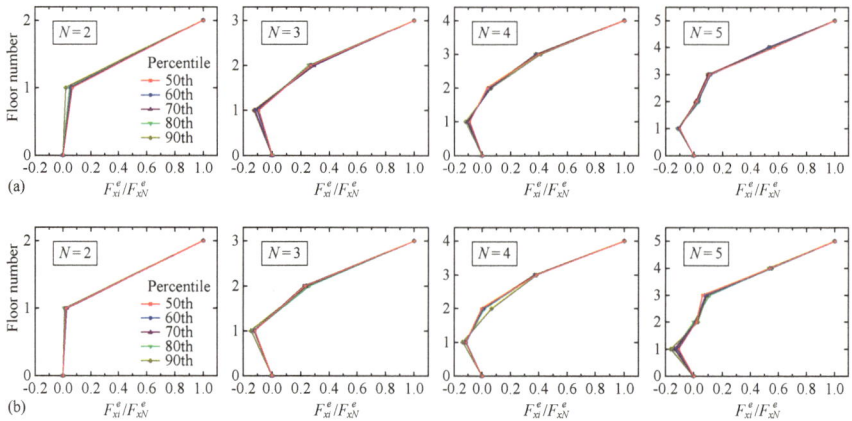

Fig. 5.4 Distributions of F_{xi}^e/F_{xN}^e under different percentiles ($\alpha = 0.8$): **a** $a_{gx0}/\mu g = 6$; **b** $a_{gx0}/\mu g = 12$

Identify the peak story shear of each story for every ground motion; (2) Organize the peak story shears for each story from the 180 ground motion records in ascending order; (3) Determine equivalent lateral forces using the peak story shears at the corresponding percentile. The equivalent lateral force of the ith floor is represented as F_{xi}^e. The distributions of F_{xi}^e/F_{xN}^e for various percentiles are illustrated in Fig. 5.4 when $\alpha = 0.8$. As shown in Fig. 5.4, the selected percentile does not have a significant impact on the distribution of F_{xi}^e/F_{xN}^e. Accordingly, the distribution of F_{xi}^e/F_{xN}^e for the 50th percentile is selected for the subsequent analyses.

5.3.1 Parametric Study

Figure 5.5 depicts the data points (F_{xi}^e/F_{xN}^e, i/N) corresponding to different N in the same plot. As shown in this figure, the trend is almost the same for the data points (F_{xi}^e/F_{xN}^e, i/N) for different N. Therefore, the same relationship between F_{xi}^e/F_{xN}^e and i/N can be used for different N.

The distributions of F_{xi}^e/F_{xN}^e for various α and $a_{gx0}/\mu g$ are shown Figs. 5.6 and 5.7. The value of F_{xi}^e/F_{xN}^e for $i < N$ decreases with an increase in α or $a_{gx0}/\mu g$; and as $a_{gx0}/\mu g$ increases, the distribution of F_{xi}^e/F_{xN}^e becomes fixed. When $a_{gx0}/\mu g$ exceeds a specific value, the corresponding lateral forces at the lower levels invert direction for $\alpha \geq 0.7$. These results means that the increase of α or $a_{gx0}/\mu g$ results in the concentration of equivalent lateral forces at the upper floors.

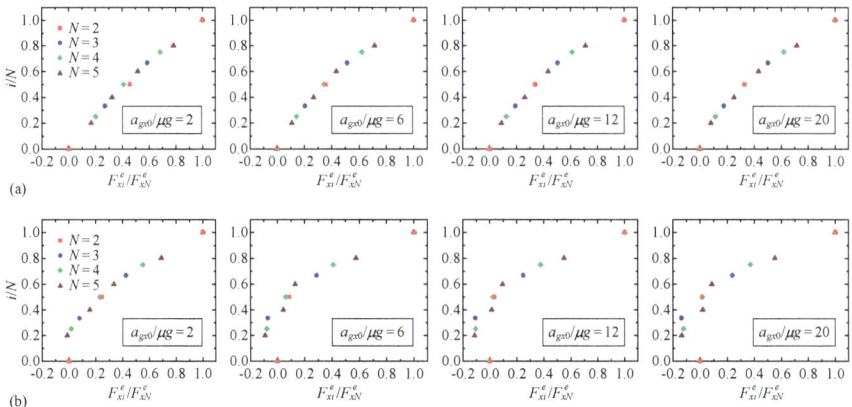

Fig. 5.5 Comparison of the relationships between F_{xi}^e / F_{xN}^e and i/N for different N: **a** $\alpha = 0.5$; **b** $\alpha = 0.8$

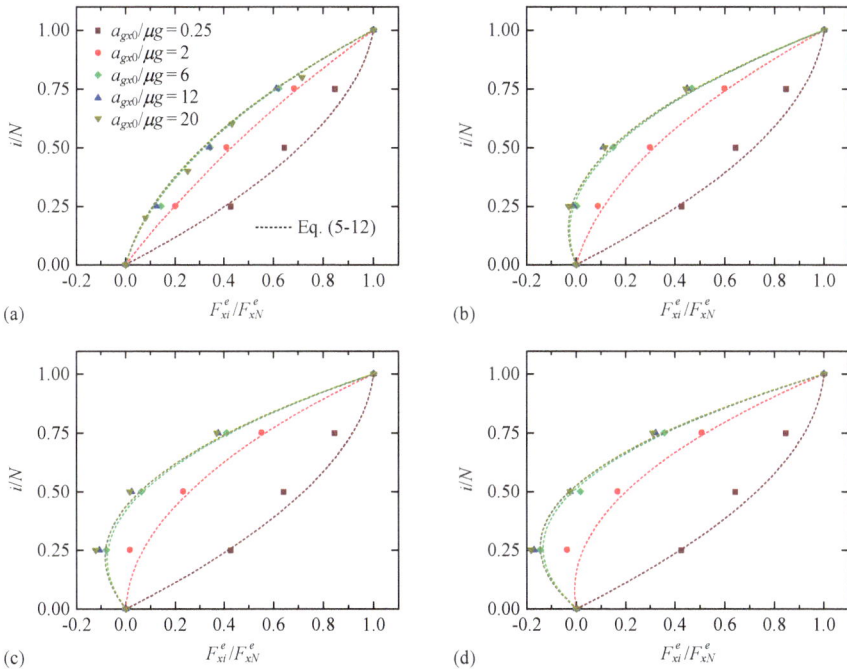

Fig. 5.6 Distributions of F_{xi}^e / F_{xN}^e for **a** $\alpha = 0.5$; **b** $\alpha = 0.7$; **c** $\alpha = 0.8$; **d** $\alpha = 0.9$ ($N = 4$)

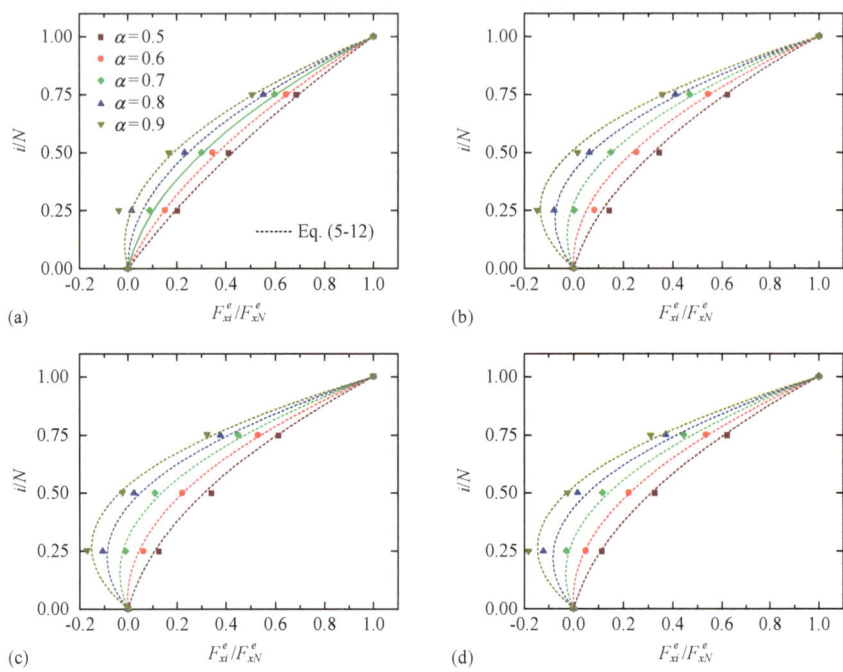

Fig. 5.7 Distributions of F_{xi}^e/F_{xN}^e for **a** $a_{gx0}/\mu g = 2$; **b** $a_{gx0}/\mu g = 6$; **c** $a_{gx0}/\mu g = 12$; and **d** $a_{gx0}/\mu g = 20$ ($N = 4$)

5.3.2 Simplified Equations

Referring to the shapes of the $i/N - F_{xi}^e/F_{xN}^e$ curves depicted in Figs. 5.6 and 5.7, the distribution of equivalent lateral forces was modeled using the following equation:

$$F_{xi}^e/F_{xN}^e = c(i/N)^2 + (1-c)(i/N) \qquad (5.12)$$

The regression coefficient c in Eq. (5.12) is dependent on $a_{gx0}/\mu g$ and α. Equation (5.12) satisfies the boundary conditions that when $i = 0$, $F_{xi}^e/F_{xN}^e = 0$, and when $i = N$, $F_{xi}^e/F_{xN}^e = 1$. The values of c attained from nonlinear regression analyses for varying values of $a_{gx0}/\mu g$ and α are presented in Fig. 5.8. The value of c increases and approaches a constant as $a_{gx0}/\mu g$ increases for a given α, and it also increases with an increase in α. These trends are in agreement with the effects of $a_{gx0}/\mu g$ and α on the distributions of F_{xi}^e/F_{xN}^e, which has been discussed in the previous section. Through nonlinear regression analyses, the following equation was developed for calculating the value of c:

$$c = (-4.3\alpha + 0.29)e^{-0.68(a_{gx0}/\mu g)} + 3.4\alpha - 0.95 \qquad (5.13)$$

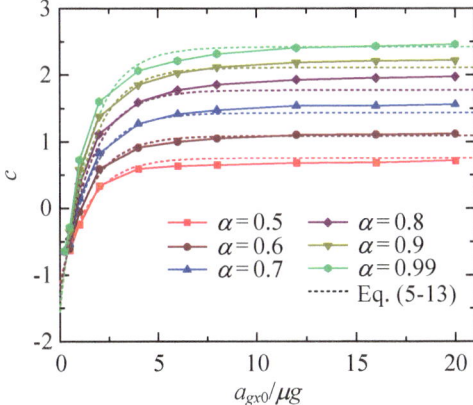

Fig. 5.8 Values of c for different values of $a_{gx0}/\mu g$ and α

As depicted in Figs. 5.6 and 5.7, the distributions of F_{xi}^{e}/F_{xN}^{e} determined using Eqs. (5.12) and (5.13) demonstrate good agreement with those that were computed based on response history analyses.

5.4 Conclusions

In this chapter, the equivalent lateral forces for the design of multistory SB structures are studied. The increase in story number N tends to reduce the mean value of the normalized peak base shear but has little effect on the coefficient of variation. The peak story shears of a multistory SB structure under an earthquake ground motion occur at varying times for each story. Based on this truth, the equivalent lateral forces can be calculated by utilizing the peak story shears that correspond to the same percentile. Based on the computed results, it can be inferred that the distribution of equivalent lateral forces is generally not influenced by the percentile chosen. The distributions of equivalent lateral forces for various N follow a similar pattern. Additionally, with an increase in the normalized PGA and mass ratio, the equivalent lateral forces tend to concentrate at the upper floors. Using Eqs. (5.9), (5.11), (5.12), and (5.13), the equivalent lateral forces required for the design of multistory SB structures can be determined.

Chapter 6
Peak Sliding Displacements of Sliding Base Structures Under Earthquake Excitation

6.1 Selection of Ground Motion Intensity Measure

The 180 ordinary ground motion records and 60 near-fault pulse-like records on site classes C and D, which has been presented in Chap. 4, will continue to be used in the response history analyses conducted in this chapter. In Chap. 3, some typical response histories of the sliding displacements are presented. The PSDs in the two principal directions (i.e., x and y directions), u_{sx0} and u_{sy0}, can be directly determined once the response histories of the sliding displacements are obtained through response history analyses. The maximum of PSDs over all the horizontal directions or the PSD with respect to the origin, u_{st0}, can also be determined by

$$u_{st0} = \max_{t} \sqrt{u_{sx}(t)^2 + u_{sy}(t)^2} \tag{6.1}$$

In design, the PSDs of interest that are required to check the sliding displacement are reliant on the boundary shape of the sliding surface. For SB masonry structures (e.g., Nanda et al., 2015; Qamaruddin et al., 1986a), bond beams are generally constructed under the masonry walls as sliding elements, and the boundary shape of the sliding surface is generally rectangular to align with the building plane. For this case, it is necessary to check the sliding displacements separately in the two principal directions, and the PSDs required in design are u_{sx0} and u_{sy0}. Generally, when using sliding isolation bearings (e.g., Jampole et al., 2016), the boundary shape of the sliding surface is circular. For this case, design requires comparison of u_{st0} with the sliding displacement threshold. The analyses that follow investigate all of u_{sx0}, u_{sy0}, and u_{st0}, covering the two cases mentioned before.

To develop dependable earthquake excitation-based prediction models for the PSDs of SB structures, it is necessary to choose a suitable ground motion intensity measure (IM) that provides relatively small record-to-record variability of the PSDs at a given IM. The PGA and PGV, which rely exclusively on the ground motion characteristics, are the most traditional measures of ground motion intensity. Although

© The Author(s) 2023
H.-S Hu, *Sliding Base Structures*,
https://doi.org/10.1007/978-981-99-5107-9_6

there have been proven several IMs that consider both ground motion and structural properties [e.g., spectral acceleration at the first-mode period of the structure (Housner, 1941) and average spectral acceleration (Eads et al., 2015)] to be more efficient for seismic response assessment of fixed base (FB) structures, the changing dynamic property of a SB structure may render them unsuitable when sliding occurs. Thus, a feasible approach to assess the PSDs of SB structures involves choosing an IM that solely relies on the ground motion characteristics and analyzing the effect of distinct structural properties in isolation.

PGA and PGV are potential IMs due to their widespread usage among researchers and engineers, as well as the availability of corresponding attenuation relationships (Villaverde, 2009). To compare the efficiency (Luco & Cornell, 2007) of different IMs (i.e., their capability to produce small variability of the PSD at a given IM), response history analyses of a SB structure with $T_x = T_y = 0.4$ s ($T_x = 2\pi/\omega_x$ and $T_y = 2\pi/\omega_y$), $\xi_x = \xi_y = 5\%$, and $\alpha = 0.7$ subjected to the three components of the 180 ordinary ground motion records were conducted. Figure 6.1 displays the calculated values of u_{sx0} and u_{st0} in relation to their respective PGAs (a_{gx0} and a_{gt0}) and PGVs (v_{gx0} and v_{gt0}), where two levels of friction coefficient μ are considered, namely $\mu = 0.1$ and 0.2. In order to maintain consistency with the definition of u_{st0}, Eqs. (6.2) and (6.3) are respectively used to compute the PGA and PGV corresponding to u_{st0}, which are the maximum values of PGA and PGV over all the horizontal directions.

$$a_{gt0} = \max_t \sqrt{\ddot{u}_{gx}(t)^2 + \ddot{u}_{gy}(t)^2} \qquad (6.2)$$

$$v_{gt0} = \max_t \sqrt{\dot{u}_{gx}(t)^2 + \dot{u}_{gy}(t)^2} \qquad (6.3)$$

The superiority of PGV over PGA is evident from Fig. 6.1. Previous researchers (e.g., Jampole et al., 2020; Ryan & Chopra, 2004) have also observed this result. To further quantify the performance of different IMs, the results shown in Fig. 6.1 were

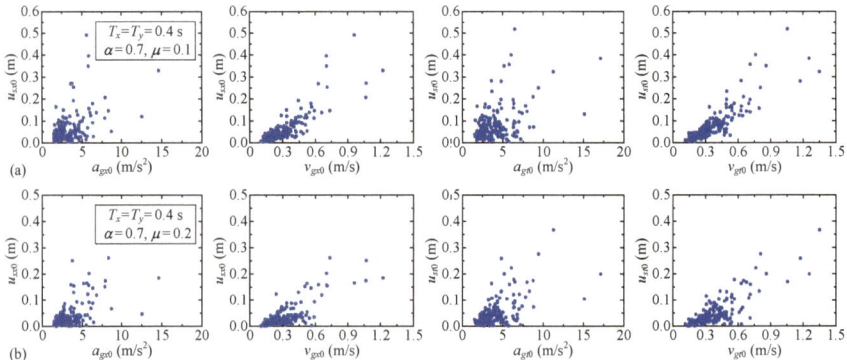

Fig. 6.1 Correlations between PSDs and corresponding PGAs and PGVs: **a** $\mu = 0.1$; and **b** $\mu = 0.2$

also subjected to correlation coefficient computation; since a linear relationship may not be the best representation of the connection between the PSD and any of the IMs under consideration, the Spearman rank correlation coefficient (Maritz, 1995), ρ_s, for nonlinear correlations is used here. For $\mu = 0.1$, the computed values of ρ_s are 0.38 (0.38) and 0.80 (0.85) for the correlations between u_{sx0} and a_{gx0} (u_{st0} and a_{gt0}), and u_{sx0} and v_{gx0} (u_{st0} and v_{gt0}), respectively; and for $\mu = 0.2$, these values are 0.56 (0.60) and 0.69 (0.73), respectively. The efficiency of PGV as an IM is relatively high, and it improves with increased sliding extent, which is appreciated because design is mainly concerned with sliding displacements that are sufficiently large and may exceed the preset threshold. Nevertheless, the variability of the PSD at a given PGV is still considerable in comparison with the peak superstructure response of SB structures presented in Chap. 4. This relatively large variability is primarily attributed to the following reasons, which have been pointed out by Jampole et al. (2020):

(1) The initiation of sliding is dominated by the acceleration quantities; thus, initiating sliding through a pulse with larger PGV may not be easier as the larger PGV could be the result of longer duration instead of a larger acceleration amplitude.

(2) Although the incremental velocity of a pulse can effectively describe the sliding excursion resulting from a velocity pulse, the incremental velocity of the largest pulse may not necessarily be in close proximity to the PGV of a ground motion record because the value of PGV is also affected by the initial conditions preceding the largest pulse.

(3) The PSD obtained from seismic excitation of a SB structure is an accumulative result of multiple sliding excursions initiated by large velocity pulses, especially for lower friction levels. The efficiency of PGV is further decreased due to the accumulative effect, as it is typically associated with a dominant pulse.

Jampole et al. (2020) suggested a new IM, named EIGV, that is more effective in forecasting PSDs of rigid bodies exposed to earthquake ground motions. However, because the correlation between the PSD and PGV is acceptable and PGV is simple and well accepted by the engineering community as a ground motion IM, PGV is adopted herein.

6.2   Critical Parameters and Their Ranges

The parametric study presented in Chap. 5 indicates that the story number N generally does not influence the distribution of the equivalent lateral forces, which implies that the sliding displacement of a multistory SB structure should be close to that of the corresponding single-story SB structure with the same mass ratio and fundamental period. To verify this inference, the responses of a three-story and a single-story SB structure with the mass ratio $\alpha = 0.75$ and the fundamental periods in the x and y directions equal to 0.4 s were computed. Figure 6.2 compares the probability

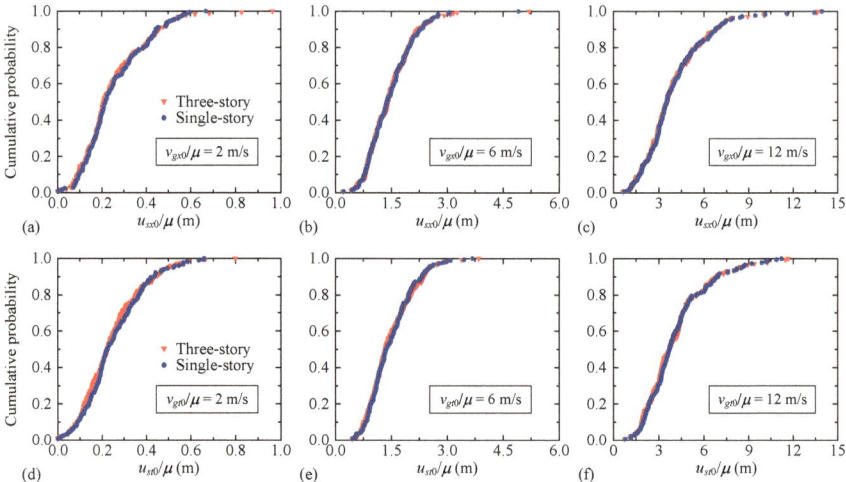

Fig. 6.2 Comparison of the probability distributions of the normalized PSDs of three-story and single-story SB structures ($\alpha = 0.75$ and $T_x = T_y = 0.4$ s): **a** $v_{gx0}/\mu = 2$ m/s; **b** $v_{gx0}/\mu = 6$ m/s; **c** $v_{gx0}/\mu = 12$ m/s; **d** $v_{gr0}/\mu = 2$ m/s; **e** $v_{gr0}/\mu = 6$ m/s; and **f** $v_{gr0}/\mu = 12$ m/s

distributions of the PSDs normalized by the friction coefficient μ of these two structures. As can be seen, the probability distributions of the normalized PSDs of the three-story and single-story structures almost coincide at a given level of the PGV normalized by μ. Therefore, single-story SB structures can be used to evaluate the PSDs of general SB structures.

As already presented in Chap. 4, the governing equations of single-story SB structures are as follows: For the stick phases,

$$
\begin{cases}
\ddot{u}_{rx} + 2\xi_x \omega_x \dot{u}_{rx} + \omega_x^2 u_{rx} = -\ddot{u}_{gx} \\
\ddot{u}_{ry} + 2\xi_y \omega_y \dot{u}_{ry} + \omega_y^2 u_{ry} = -\ddot{u}_{gy}
\end{cases}
\tag{6.4}
$$

The precondition for the stick phases is

$$
\sqrt{\left(\alpha \ddot{u}_{rx} + \ddot{u}_{gx}\right)^2 + \left(\alpha \ddot{u}_{ry} + \ddot{u}_{gy}\right)^2} < \left(g + \ddot{u}_{gz}\right)\mu_s
\tag{6.5}
$$

For the sliding phases,

$$
\begin{cases}
\ddot{u}_{sx} + \ddot{u}_{rx} + 2\xi_x \omega_x \dot{u}_{rx} + \omega_x^2 u_{rx} = -\ddot{u}_{gx} \\
\ddot{u}_{sx} + \dfrac{\dot{u}_{sx}}{\sqrt{\dot{u}_{sx}^2 + \dot{u}_{sy}^2}}\left(g + \ddot{u}_{gz}\right)\mu + \alpha \ddot{u}_{rx} = -\ddot{u}_{gx} \\
\ddot{u}_{sy} + \ddot{u}_{ry} + 2\xi_y \omega_y \dot{u}_{ry} + \omega_y^2 u_{ry} = -\ddot{u}_{gy} \\
\ddot{u}_{sy} + \dfrac{\dot{u}_{sy}}{\sqrt{\dot{u}_{sx}^2 + \dot{u}_{sy}^2}}\left(g + \ddot{u}_{gz}\right)\mu + \alpha \ddot{u}_{ry} = -\ddot{u}_{gy}
\end{cases}
\tag{6.6}
$$

As can be seen from Eqs. (6.4) to (6.6), the responses of SB structures are influenced by the friction coefficient (the dynamic and static friction coefficients are assumed to be the same), which is a critical parameter. By dividing both sides of Eqs. (6.4)–(6.6) by μ, it can be observed that the effects of μ can be incorporated into both the response quantities and ground motion IM, i.e., the normalized displacement quantities, $\overline{u}_{rx}(t) = u_{rx}(t)/\mu$, $\overline{u}_{ry}(t) = u_{ry}(t)/\mu$, $\overline{u}_{sx}(t) = u_{sx}(t)/\mu$, and $\overline{u}_{sy}(t) = u_{sy}(t)/\mu$, are dependent on the normalized IMs, v_{gx0}/μ and v_{gy0}/μ (where v_{gx0} and v_{gy0} are the PGVs in the x and y directions, respectively), and μ is not an independent variable anymore. Thus, to simplify the estimation of PSDs related to different levels of ground motion intensity associated with various levels of μ, we can evaluate the normalized PSDs (u_{sx0}/μ, u_{sy0}/μ, and u_{st0}/μ) at various normalized PGVs (v_{gx0}/μ, v_{gy0}/μ, and v_{gt0}/μ). Equations (6.4)–(6.6) shows that ω_x, ω_y, ξ_x, ξ_y, α, $\ddot{u}_{gz}(t)$, and the horizontal ground motion waveform are other parameters that may affect the normalized PSDs.

The common ranges of the natural periods of the superstructure ($T_x = 2\pi/\omega_x$ and $T_y = 2\pi/\omega_y$), the damping ratios (ξ_x and ξ_y), and the mass ratio (α) have been presented in Sect. 4.2, and thus are not repeated here. The friction coefficients of the sliding interfaces investigated for SB structures (Barbagallo et al., 2017; Hasani, 1996; Jampole et al., 2016; Nanda et al., 2012; Qamaruddin et al., 1986; Yegian et al., 2004) range from 0.07 to 0.41. Apart from very few near-fault records from high-magnitude earthquakes, most of the ground motions recorded have PGVs below 1.2 m/s. On these bases, the normalized PGV is limited to 18 m/s, which is equivalent to PGV = 1.26 m/s when $\mu = 0.07$. When the normalized PGV is 1 m/s, the PSDs associated with μ lying in the common range are well below 0.1 m, a value that can be considered as the lower bound of the sliding displacement threshold in practice. Therefore, analyzing the cases with normalized PGV below 1 m/s is not necessary from a design perspective. In the following parametric study, eleven levels of normalized PGVs, namely 1, 1.5, 2, 4, 6, 8, 10 12, 14, 16, and 18 m/s, are considered. These levels of normalized PGVs were achieved by adjusting the value of μ with the ground motion records unscaled.

6.3 Parametric Study for the Normalized Peak Sliding Displacements

6.3.1 Comparison of the Responses in the Two Orthogonal Directions

The coupling of the friction forces in the two orthogonal directions [i.e., their resultant is equal to $(m + m_b)(g + \ddot{u}_{gz})\mu$] causing the ground motion in one direction tends to decrease the friction force component in the orthogonal direction, resulting in an increase in the sliding displacement in that direction. However, this effect is reciprocal. Assuming using enough number of ground motions, the relationship between

the average normalized PSD and normalized PGV in one direction that was obtained
should be the same as that in the orthogonal direction under the circumstance of
three-component seismic excitation. To confirm this inference, Fig. 6.3 displays the
mean values of u_{sx0}/μ at each level of v_{gx0}/μ considered and the data points, $(v_{gy0}/\mu,$
$u_{sy0}/\mu)$, corresponding to the response in the y direction. These data were obtained
from response history analyses of SB structures with $T_x = T_y = 0.4$ s and $\alpha = 0.7$
using the 180 non-pulse-like ground motion records. Figure 6.3 presents only the
data points with 1 m/s $\le v_{gy0}/\mu \le 18$ m/s, in order to maintain consistency with
the range of the normalized ground motion IM considered in the x direction. It was
found that a quadratic polynomial curve can well represent the relationship between
mean u_{sx0}/μ and v_{gx0}/μ; therefore, a regression curve, obtained through the use of
a quadratic polynomial equation for fitting the data points, is displayed in Fig. 6.3.
As can be seen in this figure, the regression curve for the relationship between the
mean u_{sy0}/μ and v_{gy0}/μ agrees well with the curve of the mean u_{sx0}/μ versus v_{gx0}/μ.
Therefore, it can be inferred that the relationship between the mean normalized PSD
and normalized PGV is essentially identical for the two orthogonal horizontal directions. However, it is important to note that the PSDs may vary greatly between the
two orthogonal directions for an individual ground motion, despite both directions
having the same PGVs. In terms of statistical results, the outcomes achieved for the
x-direction through a considerable number of ground motions can be extended to
the y-direction. For this reason, only the response in the x direction is investigated
hereafter.

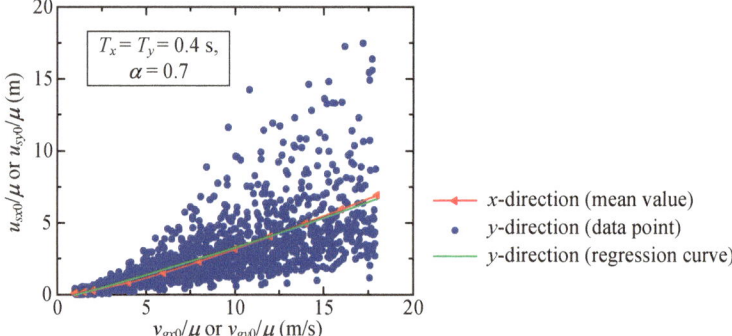

Fig. 6.3 Comparison of the relationship between the normalized PSD and normalized PGV for the
two orthogonal horizontal directions

6.3.2 Probability Distribution of the Normalized PSD at a Given Level of Normalized Ground Motion Intensity

Figure 6.4 depicts the cumulative probability distribution of the normalized PSD (u_{sx0}/μ and u_{st0}/μ) at four distinct levels of normalized PGV (v_{gx0}/μ and v_{gt0}/μ). The data used to determine these empirical cumulative distributions were derived from response history analyses using the 180 non-pulse-like ground motion records with the structural parameters $T_x = T_y = 0.4$ s and $\alpha = 0.7$. The figure clearly depicts that the empirical distributions are asymmetrical around the sample median and have lengthier tails moving towards upper values. The lognormal distribution, which has been extensively utilized in seismic performance assessment of building structures (e.g., Ruiz-Garcia & Miranda, 2007; Zareian & Krawinkler, 2007), also presents such type of feature and, thus, could be suitable for modeling the probability distributions of u_{sx0}/μ at a given level of v_{gx0}/μ and u_{st0}/μ at a given level of v_{gt0}/μ. The sample geometric mean and sample logarithmic standard deviation (Ang & Tang, 2006) are typically used to estimate the two parameters (i.e., the median and logarithmic standard deviation) of the fitted lognormal distribution function. For this study, the equations for estimating the parameters can be written as

$$(u_{sx0}/\mu)_m = \exp\left(\sum_{i=1}^{n} \ln(u_{sx0}/\mu)_i / n\right) \tag{6.7a}$$

$$(u_{st0}/\mu)_m = \exp\left(\sum_{i=1}^{n} \ln(u_{st0}/\mu)_i / n\right) \tag{6.7b}$$

$$\sigma_{\ln(u_{sx0}/\mu)} = \sqrt{\frac{\sum_{i=1}^{n} \left[\ln(u_{sx0}/\mu)_i - \ln(u_{sx0}/\mu)_m\right]^2}{n-1}} \tag{6.8a}$$

$$\sigma_{\ln(u_{st0}/\mu)} = \sqrt{\frac{\sum_{i=1}^{n} \left[\ln(u_{st0}/\mu)_i - \ln(u_{st0}/\mu)_m\right]^2}{n-1}} \tag{6.8b}$$

where $(u_{sx0}/\mu)_m$ and $(u_{st0}/\mu)_m$ are the medians of u_{sx0}/μ and u_{st0}/μ, respectively; $\sigma_{\ln(u_{sx0}/\mu)}$ and $\sigma_{\ln(u_{st0}/\mu)}$ are the lognormal standard deviations of u_{sx0}/μ and u_{st0}/μ, respectively; $(u_{sx0}/\mu)_i$ and $(u_{st0}/\mu)_i$ are the observed value; and n is the sample size. However, for v_{gx0}/μ (and v_{gt0}/μ) ≤ 2 m/s, some observed values of u_{sx0}/μ (and u_{st0}/μ) are 0 or very close to 0, which makes Eqs. (6.7) and (6.8) invalid because the natural logarithm of zero does not exist and the value computed by Eq. (6.7) will be dominated by the natural logarithm of a value that is near 0. Thus, for these cases, $(u_{sx0}/\mu)_m$ and $(u_{st0}/\mu)_m$ are taken as the counted medians, and $\sigma_{\ln(u_{sx0}/\mu)}$ and $\sigma_{\ln(u_{st0}/\mu)}$ are estimated by Eq. (6.9) based on the assumption that the data are sampled from lognormal distributions.

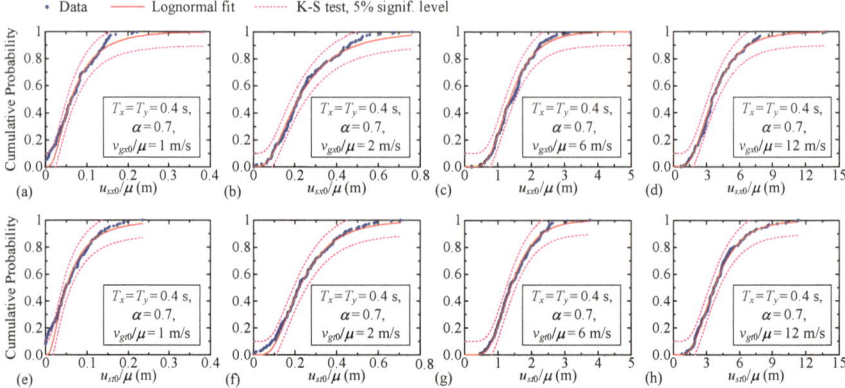

Fig. 6.4 Empirical and fitted lognormal probability distributions of the normalized PSD at given levels of corresponding normalized PGV: **a** $v_{gx0}/\mu = 1$ m/s; **b** $v_{gx0}/\mu = 2$ m/s; **c** $v_{gx0}/\mu = 6$ m/s; **d** $v_{gx0}/\mu = 12$ m/s; **e** $v_{gt0}/\mu = 1$ m/s; **f** $v_{gt0}/\mu = 2$ m/s; **g** $v_{gt0}/\mu = 6$ m/s; **h** $v_{gt0}/\mu = 12$ m/s

$$\sigma_{\ln(u_{sx0}/\mu)} = \ln\left[\frac{(u_{sx0}/\mu)_{84\%}}{(u_{sx0}/\mu)_{50\%}}\right] \tag{6.9a}$$

$$\sigma_{\ln(u_{st0}/\mu)} = \ln\left[\frac{(u_{st0}/\mu)_{84\%}}{(u_{st0}/\mu)_{50\%}}\right] \tag{6.9b}$$

where $(u_{sx0}/\mu)_{50\%}$ and $(u_{sx0}/\mu)_{84\%}$ are the counted median and counted 84th percentile of u_{sx0}/μ, respectively; and $(u_{st0}/\mu)_{50\%}$ and $(u_{st0}/\mu)_{84\%}$ are the counted median and counted 84th percentile of u_{st0}/μ, respectively.

Figure 6.4 also displays the fitted lognormal distribution functions for each of the four levels of v_{gx0}/μ and v_{gt0}/μ. In general, the fitted lognormal distribution agrees fairly well with the corresponding empirical distribution. The well-known Kolmogorov–Smirnov (K-S) goodness-of-fit tests (Ang & Tang, 2006) were conducted to further verify the adequacy of the lognormal distribution. Figure 6.4 depicts the graphical representations of the K-S test with a 5% significance level. The figure displays that all data points for v_{gx0}/μ (and v_{gt0}/μ) = 2, 6, and 12 m/s, are within the limits of acceptability (i.e., the two dotted lines in Fig. 6.4), indicating that the assumed lognormal distribution is acceptable. For v_{gx0}/μ (and v_{gt0}/μ) = 1 m/s, due to the presence of several null values, certain points at the lower tail fall outside the acceptable limits; however, the practical sliding displacement threshold is much higher than the PSDs associated with this lower tail, thus the utility of the lognormal distribution remains unaffected. Concluding from the aforementioned discussions, it is evident that the lognormal distribution is appropriate for modeling the probability distributions of u_{sx0}/μ at a given level of v_{gx0}/μ, and u_{st0}/μ at a given level of v_{gt0}/μ.

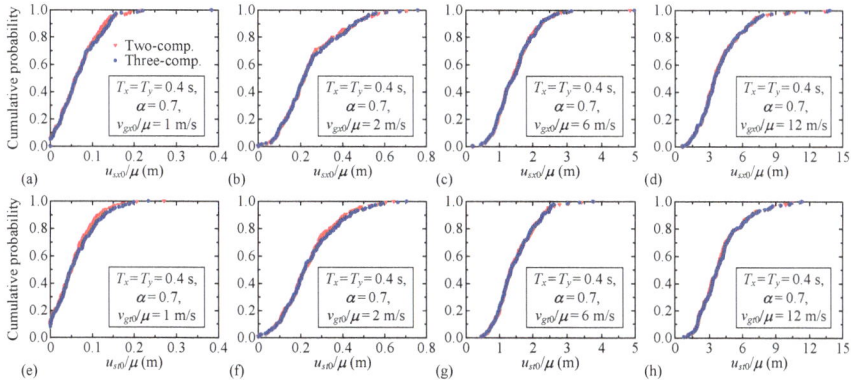

Fig. 6.5 Comparison of the probability distributions of the normalized PSD under two- and three-component excitations: **a** $v_{gx0}/\mu = 1$ m/s; **b** $v_{gx0}/\mu = 2$ m/s; **c** $v_{gx0}/\mu = 6$ m/s; **d** $v_{gx0}/\mu = 12$ m/s; **e** $v_{gt0}/\mu = 1$ m/s; **f** $v_{gt0}/\mu = 2$ m/s; **g** $v_{gt0}/\mu = 6$ m/s; **h** $v_{gt0}/\mu = 12$ m/s

6.3.3 Effect of the Vertical Ground Motion Component

To investigate the influence of the vertical component on the PSD, the computation was carried out on the responses of SB structures exposed solely to the two horizontal components of the 180 non-pulse-like ground motions. Figure 6.5 shows a comparison between the normalized PSD probability distribution under the two-component excitation and that of corresponding three-component excitation. The normalized PSD probability distributions for both cases in this figure are almost the same at a given normalized PGV level, with only slight differences in the upper portions of the cumulative distribution curves. This result indicates that the effect of the vertical component on the PSD is negligible. Shao and Tung (1999) and Konstantinidis and Nikfar (2015) have also arrived at comparable conclusions regarding the sliding behavior of rigid bodies.

6.3.4 Effects of the Superstructure Natural Period and Mass Ratio

Figures 6.6 and 6.7 present the relationships between $(u_{sx0}/\mu)_m$ and v_{gx0}/μ, and $(u_{st0}/\mu)_m$ and v_{gt0}/μ, respectively, for different values of T_x and α, which were determined by using the 180 non-pulse-like ground motion records and assuming $T_x = T_y$. As shown in these figures, the trend of $(u_{sx0}/\mu)_m$ changing as v_{gx0}/μ increases closely resembles the trend of $(u_{st0}/\mu)_m$ changing as v_{gt0}/μ increases. In comparison with the normalized PGV, the influence of T_x and α on $(u_{sx0}/\mu)_m$ and $(u_{st0}/\mu)_m$ is not so significant. To further investigate the combined effects of T_x and α on $(u_{sx0}/\mu)_m$ and $(u_{st0}/\mu)_m$, $(u_{sx0}/\mu)_m$ and $(u_{st0}/\mu)_m$ are plotted against T_x and α in Figs. 6.8 and

6.9, respectively, for four representative levels of normalized PGV. This figure also presents the results of rigid bodies, which correspond to $T_x = 0$. As can be seen in Figs. 6.8 and 6.9, when $T_x \leq 0.4$ s, the mass ratio basically has a negligible effect on $(u_{sx0}/\mu)_m$ and $(u_{st0}/\mu)_m$; when $T_x > 0.4$ s, the influence of α becomes slightly more significant, and $(u_{sx0}/\mu)_m$ and $(u_{st0}/\mu)_m$ generally increase as α increases. This phenomenon cannot be simply interpreted using the governing equations presented previously; additionally, since a larger value of α does not always lead to a larger $(u_{sx0}/\mu)_m$ or $(u_{st0}/\mu)_m$, as presented in Figs. 6.8 and 6.9, the inherent characteristics of the ground motion time history should have played a significant role in this general trend. For a given mass ratio, the values of $(u_{sx0}/\mu)_m$ and $(u_{st0}/\mu)_m$ generally first increase and then decrease as T_x increases, and the differences between the maximum and minimum values of $(u_{sx0}/\mu)_m$ and $(u_{st0}/\mu)_m$ for T_x within the range considered range from 0.04 to 0.33 m and 0.02 to 0.27 m, respectively, and generally increase as the corresponding normalized PGV increases. From this result, we know that the PSDs of actual SB structures may be underestimated by relying solely on the response of rigid bodies. For simplicity, it is reasonable to use the maximum values of $(u_{sx0}/\mu)_m$ and $(u_{st0}/\mu)_m$ for the range of T_x considered to conservatively estimate the PSDs of possible SB structures.

Figures 6.10 and 6.11 present the relationships between $\sigma_{\ln(u_{sx0}/\mu)}$ and v_{gx0}/μ, and $\sigma_{\ln(u_{st0}/\mu)}$ and v_{gt0}/μ, respectively, for different values of T_x and α. For some cases when v_{gx0}/μ (and v_{gt0}/μ) = 1 m/s, the values of $(u_{sx0}/\mu)_{50\%}$ [and $(u_{st0}/\mu)_{50\%}$] are 0 or very close to 0; thus, the values obtained from using Eq. (6.9) to compute $\sigma_{\ln(u_{sx0}/\mu)}$ [and $\sigma_{\ln(u_{st0}/\mu)}$] are infinite or unreasonably large. For this reason, the results corresponding to v_{gx0}/μ (and v_{gt0}/μ) = 1 m/s are not presented in Fig. 6.10 (and Fig. 6.11). As shown in Fig. 6.10 (and Fig. 6.11), $\sigma_{\ln(u_{sx0}/\mu)}$ [and $\sigma_{\ln(u_{st0}/\mu)}$] generally

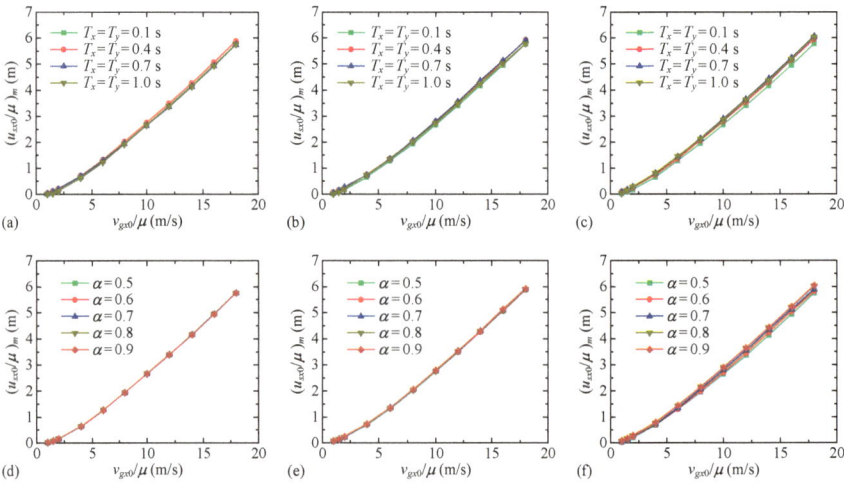

Fig. 6.6 Relationships between $(u_{sx0}/\mu)_m$ and v_{gx0}/μ for different values of T_x and α: **a** $\alpha = 0.5$; **b** $\alpha = 0.7$; **c** $\alpha = 0.9$; **d** $T_x = 0.1$ s; **e** $T_x = 0.4$ s; **f** $T_x = 0.7$ s

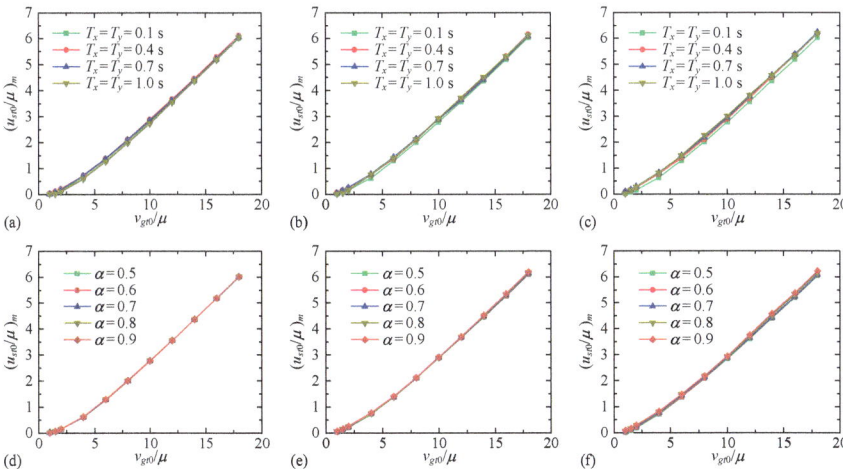

Fig. 6.7 Relationships between $(u_{st0}/\mu)_m$ and v_{gt0}/μ for different values of T_x and α: **a** $\alpha = 0.5$; **b** $\alpha = 0.7$; **c** $\alpha = 0.9$; **d** $T_x = 0.1$ s; **e** $T_x = 0.4$ s; **f** $T_x = 0.7$ s

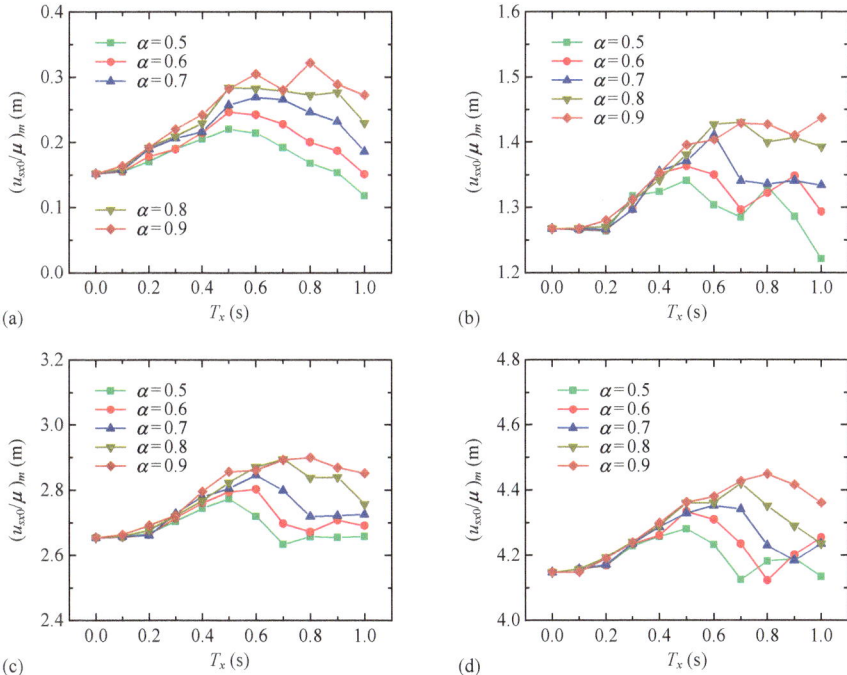

Fig. 6.8 Combined effects of T_x and α on $(u_{sx0}/\mu)_m$: **a** $v_{gx0}/\mu = 2$ m/s; **b** $v_{gx0}/\mu = 6$ m/s; **c** $v_{gx0}/\mu = 10$ m/s; **d** $v_{gx0}/\mu = 14$ m/s

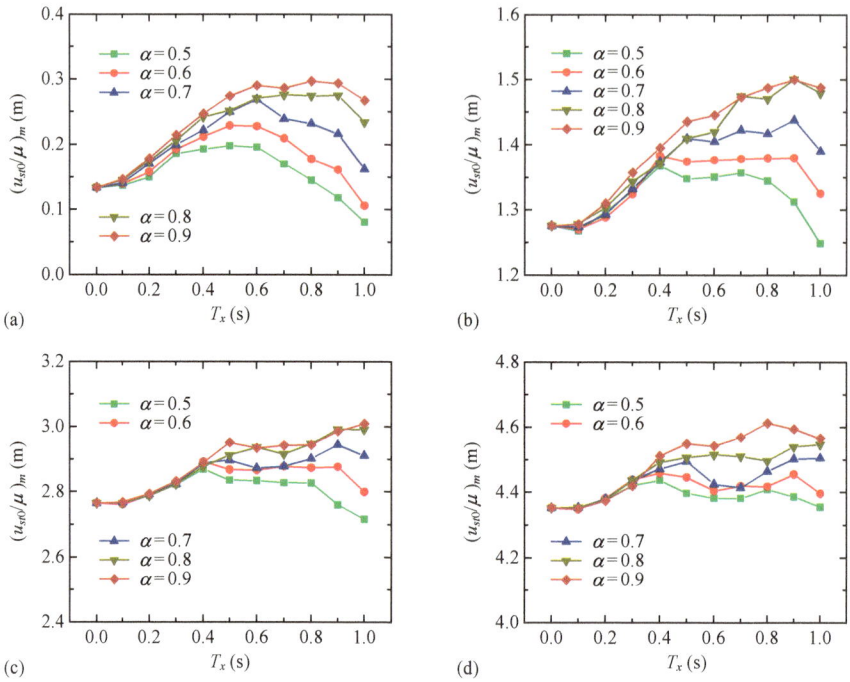

Fig. 6.9 Combined effects of T_x and α on $(u_{st0}/\mu)_m$: **a** $v_{gt0}/\mu = 2$ m/s; **b** $v_{gt0}/\mu = 6$ m/s; **c** $v_{gt0}/\mu = 10$ m/s; **d** $v_{gt0}/\mu = 14$ m/s

lie between 0.4 and 0.6 except for some cases when v_{gx0}/μ (and v_{gt0}/μ) $= 1.5$ and 2 m/s. When the normalized PGV is small, sliding is not predominant, and the ground acceleration may play a more significant role than the ground velocity as the acceleration quantities dominate the initiation of sliding as revealed by Eq. (6.5); furthermore, sliding is less likely to occur for smaller values of α and larger values of T_x, as demonstrated in Chap. 4. This explains why the values of $\sigma_{\ln(u_{sx0}/\mu)}$ [and $\sigma_{\ln(u_{st0}/\mu)}$] are generally larger for v_{gx0}/μ (and v_{gt0}/μ) $= 1.5$ and 2 m/s and even larger values are obtained when $T_x = 1$ s and $\alpha \leq 0.7$.

The combined effects of T_x and α on $\sigma_{\ln(u_{sx0}/\mu)}$ and $\sigma_{\ln(u_{st0}/\mu)}$ are plotted in Figs. 6.12 and 6.13, respectively, for four representative levels of normalized PGV. In general, the influence of α on $\sigma_{\ln(u_{sx0}/\mu)}$ [and $\sigma_{\ln(u_{st0}/\mu)}$] is small except for v_{gx0}/μ (and $v_{gt0}/\mu) \leq 2$ m/s. For any given level of v_{gx0}/μ (and v_{gt0}/μ), the maximum value of $\sigma_{\ln(u_{sx0}/\mu)}$ [and $\sigma_{\ln(u_{st0}/\mu)}$] is obtained at $T_x = 1$ s; this value is 0.87 (and 1.12) for v_{gx0}/μ (and $v_{gt0}/\mu) = 2$ m/s and is around 0.63 (and 0.58) for all other levels of v_{gx0}/μ (and $v_{gt0}/\mu) \geq 4$ m/s. The value of T_x at which the minimum $\sigma_{\ln(u_{sx0}/\mu)}$ [and $\sigma_{\ln(u_{st0}/\mu)}$] is obtained varies for different levels of v_{gx0}/μ (and v_{gt0}/μ). For a given level of v_{gx0}/μ (and v_{gt0}/μ), the minimum value of $\sigma_{\ln(u_{sx0}/\mu)}$ [and $\sigma_{\ln(u_{st0}/\mu)}$] ranges from 0.41 to 0.57 (and 0.38–0.53). The average value of $\sigma_{\ln(u_{sx0}/\mu)}$ [and $\sigma_{\ln(u_{st0}/\mu)}$] is

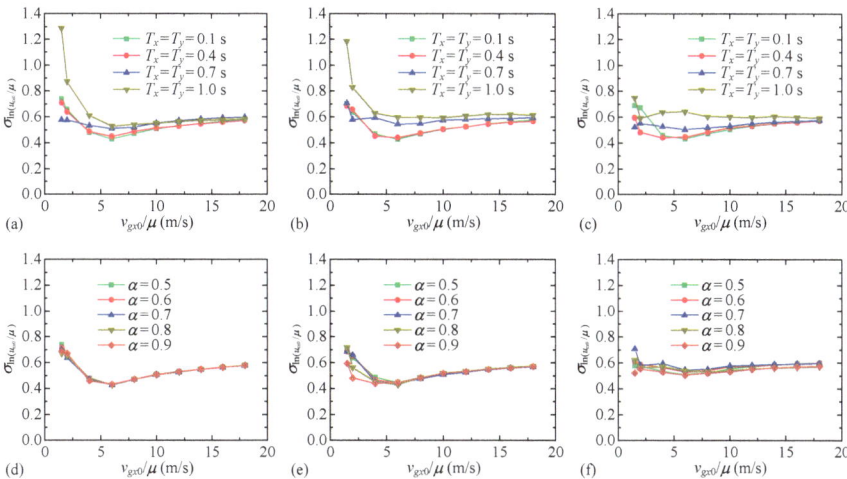

Fig. 6.10 Relationships between $\sigma_{\ln(u_{sx0}/\mu)}$ and v_{gx0}/μ for different values of T_x and α: **a** $\alpha = 0.5$; **b** $\alpha = 0.7$; **c** $\alpha = 0.9$; **d** $T_x = 0.1$ s; **e** $T_x = 0.4$ s; **f** $T_x = 0.7$ s

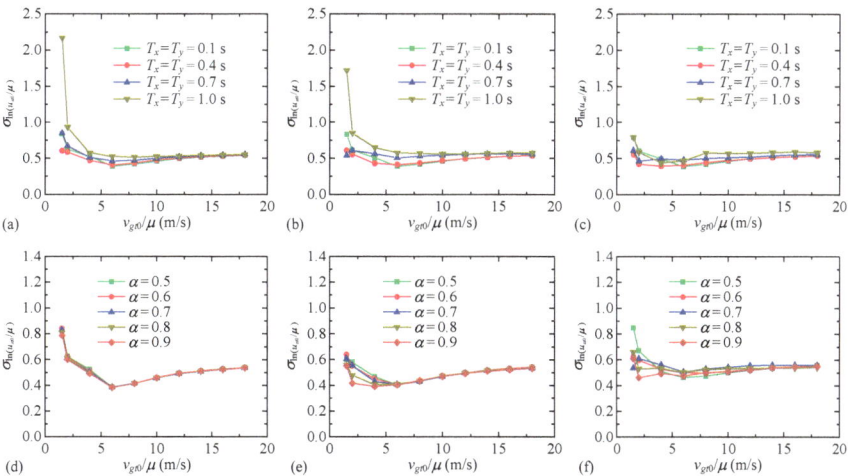

Fig. 6.11 Relationships between $\sigma_{\ln(u_{st0}/\mu)}$ and v_{gt0}/μ for different values of T_x and α: **a** $\alpha = 0.5$; **b** $\alpha = 0.7$; **c** $\alpha = 0.9$; **d** $T_x = 0.1$ s; **e** $T_x = 0.4$ s; **f** $T_x = 0.7$ s

0.60 (and 0.58) for v_{gx0}/μ (and v_{gt0}/μ) = 2 m/s and ranges from 0.49 to 0.58 (and 0.45–0.55) for v_{gx0}/μ (and v_{gt0}/μ) ≥ 4 m/s.

The equality $T_x = T_y$ is employed in all of the aforementioned analyses. To investigate the possible effect of T_x/T_y on the PSD, the values of $(u_{sx0}/\mu)_m$ and $(u_{st0}/\mu)_m$ corresponding to different values of T_x/T_y are compared in Figs. 6.14 and 6.15, respectively, and those of $\sigma_{\ln(u_{sx0}/\mu)}$ and $\sigma_{\ln(u_{st0}/\mu)}$ are compared in Figs. 6.16 and 6.17, respectively. For the data presented in these figures, α is taken as 0.7. These figures

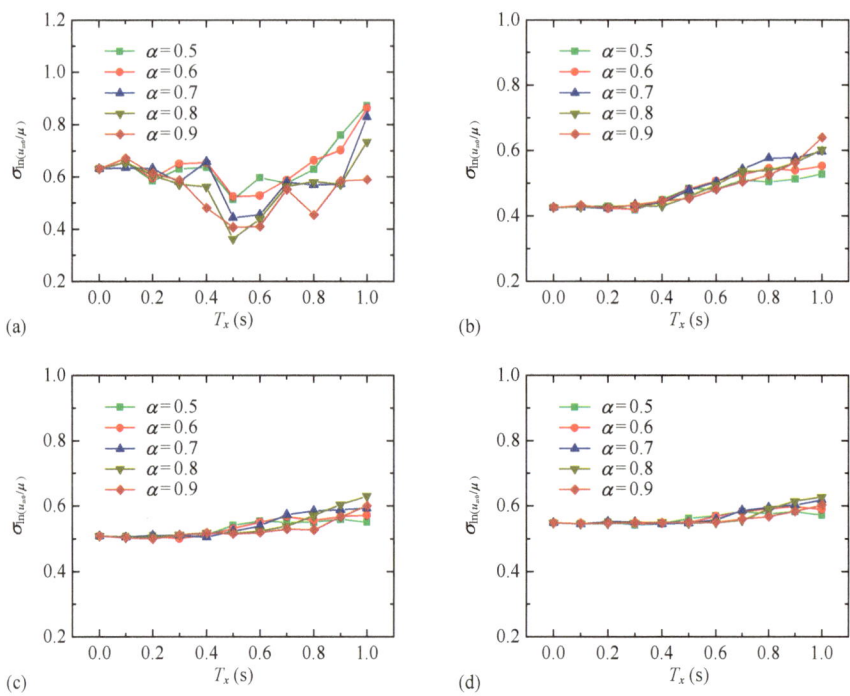

Fig. 6.12 Combined effects of T_x and α on $\sigma_{\ln(u_{sx0}/\mu)}$: **a** $v_{gx0}/\mu = 2$ m/s; **b** $v_{gx0}/\mu = 6$ m/s; **c** $v_{gx0}/\mu = 10$ m/s; **d** $v_{gx0}/\mu = 14$ m/s

make it clear that T_x/T_y has a negligible influence. Therefore, the results obtained for $T_x = T_y$ can represent those of the other T_x/T_y within the range considered.

6.3.5 Effect of Near-Fault Pulses

Distinct pulses in near-fault ground motions affected by forward directivity may result in distinct sliding response characteristics as compared to ordinary ground motions. To investigate this effect, the $(u_{sx0}/\mu)_m$ versus v_{gx0}/μ and $(u_{st0}/\mu)_m$ versus v_{gt0}/μ curves obtained using the 60 near-fault pulse-like records and the 180 non-pulse-like records are compared in Figs. 6.18 and 6.19, respectively. When $v_{gx0}/\mu \leq 4$ m/s (and $v_{gt0}/\mu \leq 6$ m/s), the values of $(u_{sx0}/\mu)_m$ [and $(u_{st0}/\mu)_m$] corresponding to the pulse-like records are close to those corresponding to the non-pulse-like records. When v_{gx0}/μ exceeds 6 m/s (and v_{gt0}/μ exceeds 8 m/s), the value of $(u_{sx0}/\mu)_m$ [and $(u_{st0}/\mu)_m$] for the pulse-like records starts to exceed the corresponding value for the non-pulse-like records, and the difference increases monotonically as v_{gx0}/μ (and v_{gt0}/μ) increases. To interpret the underlying reason for this phenomenon, Fig. 6.20 ($T_x = T_y = 0.4$ s and $\alpha = 0.7$ are adopted) illustrates the ground acceleration, velocity, and

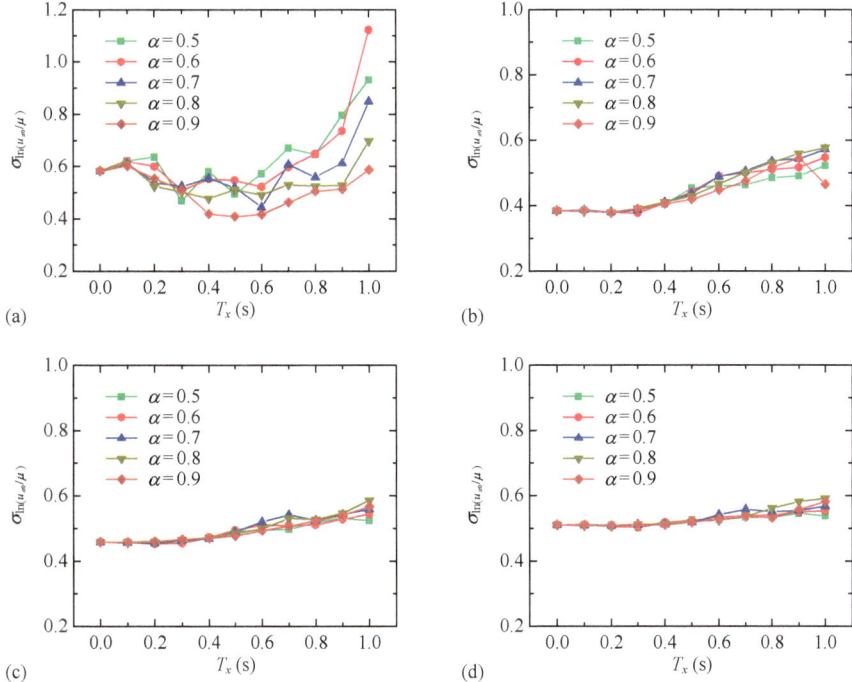

Fig. 6.13 Combined effects of T_x and α on $\sigma_{\ln(u_{st0}/\mu)}$: **a** $v_{gt0}/\mu = 2$ m/s; **b** $v_{gt0}/\mu = 6$ m/s; **c** $v_{gt0}/\mu = 10$ m/s; **d** $v_{gt0}/\mu = 14$ m/s

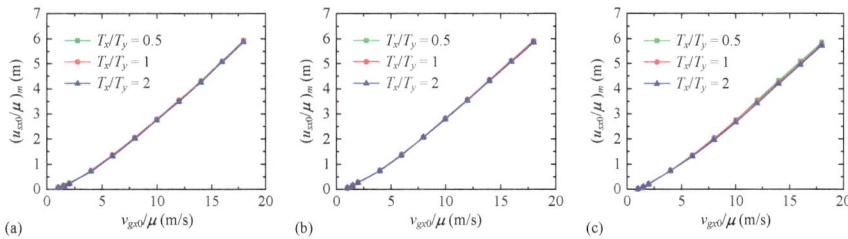

Fig. 6.14 Effects of T_x/T_y on $(u_{sx0}/\mu)_m$ ($\alpha = 0.7$): **a** $T_x = 0.4$ s; **b** $T_x = 0.7$ s; **c** $T_x = 1$ s

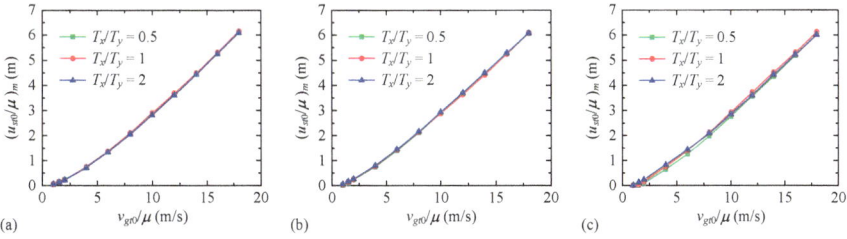

Fig. 6.15 Effects of T_x/T_y on $(u_{st0}/\mu)_m$ ($\alpha = 0.7$): **a** $T_x = 0.4$ s; **b** $T_x = 0.7$ s; **c** $T_x = 1$ s

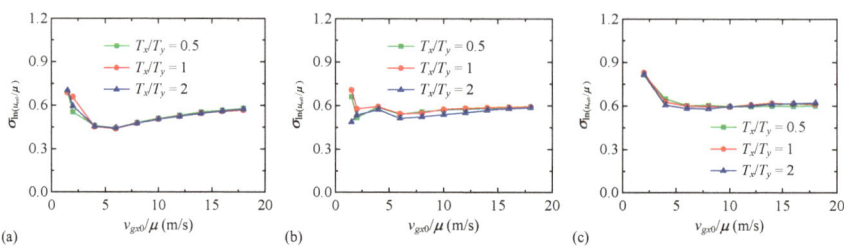

Fig. 6.16 Effects of T_x/T_y on $\sigma_{\ln(u_{sx0}/\mu)}$ ($\alpha = 0.7$): **a** $T_x = 0.4$ s; **b** $T_x = 0.7$ s; **c** $T_x = 1$ s

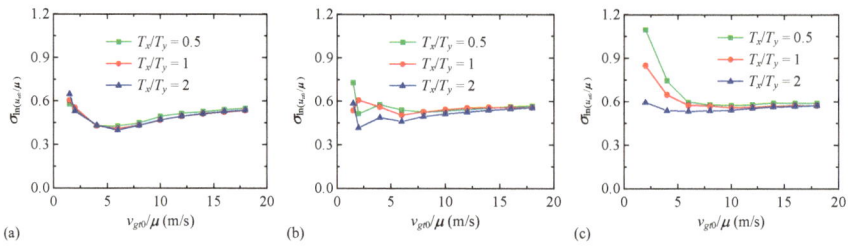

Fig. 6.17 Effects of T_x/T_y on $\sigma_{\ln(u_{st0}/\mu)}$ ($\alpha = 0.7$): **a** $T_x = 0.4$ s; **b** $T_x = 0.7$ s; **c** $T_x = 1$ s

sliding displacement time histories corresponding to the counted median of u_{sx0}/μ in each (non-pulse-like or pulse-like) group. By comparing the time histories presented in Fig. 6.20b for $v_{gx0}/\mu = 10$ m/s, we can find that the prominent long-period velocity pulse in the pulse-like ground motion is the cause of the larger value of $(u_{sx0}/\mu)_m$ in comparison with the non-pulse-like ground motion. However, when v_{gx0}/μ is small, as illustrated in Fig. 6.20a for $v_{gx0}/\mu = 2$ m/s, the contribution of the long-period velocity pulse is not so significant. The simplified equation (Eq. 6.10) proposed by Jampole et al. (2018) for predicting the PSD of a rigid block subjected to a half-sine pulse which was derived from simplification of the corresponding closed-form solution can provide an approximate interpretation of this result.

$$u_{s,\max} = \frac{a_p^2 T_p^2}{4\mu g} - \frac{1}{2}a_p T_p^2 + \frac{1}{4}T_p^2 \mu g \tag{6.10}$$

where $u_{s,\max}$ is the PSD of the rigid block; and a_p and T_p are the peak acceleration and duration of the half-sine pulse, respectively. Dividing both sides of Eq. (6.10) by μ and replacing $a_p T_p$ with $\pi v_{gi}/2$ (where v_{pi} is the peak velocity of the half-sine pulse), lead to

$$\frac{u_{s,\max}}{\mu} = \frac{\pi^2}{16g}\left(\frac{v_{pi}}{\mu}\right)^2 - \frac{\pi}{4}\frac{v_{pi}}{\mu}T_p + \frac{1}{4}T_p^2 g \tag{6.11}$$

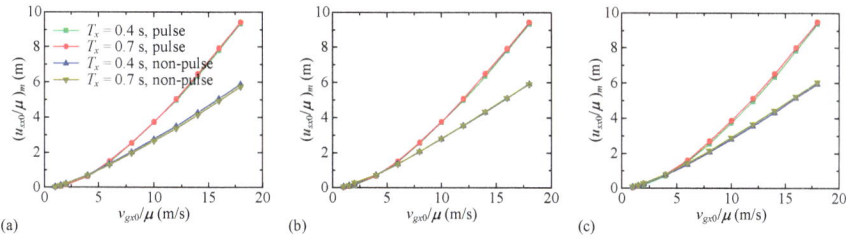

Fig. 6.18 Effects of near-fault pulses on $(u_{sx0}/\mu)_m$: **a** $\alpha = 0.5$; **b** $\alpha = 0.7$; and **c** $\alpha = 0.9$

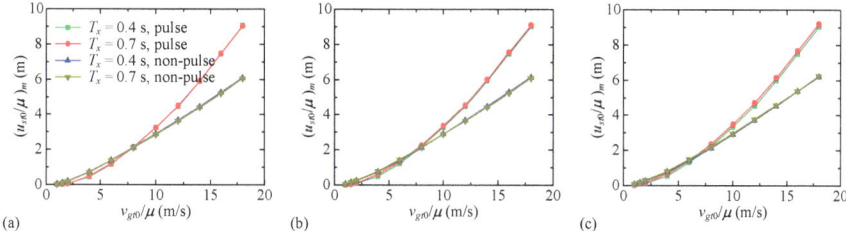

Fig. 6.19 Effects of near-fault pulses on $(u_{st0}/\mu)_m$: **a** $\alpha = 0.5$; **b** $\alpha = 0.7$; and **c** $\alpha = 0.9$

Equation (6.11) indicates that the quadratic relationship between normalized PSD and normalized PGV remains when a half-sine pulse is applied for excitation, that is, as the normalized PGV increases, the rate of increase of the normalized PSD with respect to it also increases, i.e., the effect of the prominent long-period velocity pulse is more significant when normalized PGV levels are higher.

Figures 6.21 and 6.22 compare the $\sigma_{\ln(u_{sx0}/\mu)}$ versus v_{gx0}/μ and $\sigma_{\ln(u_{st0}/\mu)}$ versus v_{gt0}/μ curves, respectively, of the pulse-like records with those of the non-pulse-like records. As shown in this figure, when v_{gx0}/μ (and v_{gt0}/μ) ≤ 4 m/s, the values of $\sigma_{\ln(u_{sx0}/\mu)}$ [and $\sigma_{\ln(u_{st0}/\mu)}$] for the pulse-like records are generally larger than those for the non-pulse-like records; when v_{gx0}/μ (and v_{gt0}/μ) ≥ 6 m/s, the value of $\sigma_{\ln(u_{sx0}/\mu)}$ [and $\sigma_{\ln(u_{st0}/\mu)}$] for the pulse-like records does not change much and is slightly smaller than the corresponding value for the non-pulse-like records. Since the computed dispersion is partly influenced by the selected ground motion records, and typically there are minimal differences in the computed $\sigma_{\ln(u_{sx0}/\mu)}$ [and $\sigma_{\ln(u_{st0}/\mu)}$] of the two ground motion types, it is reasonable to expect a similar level of inherent dispersion for the two types of ground motions.

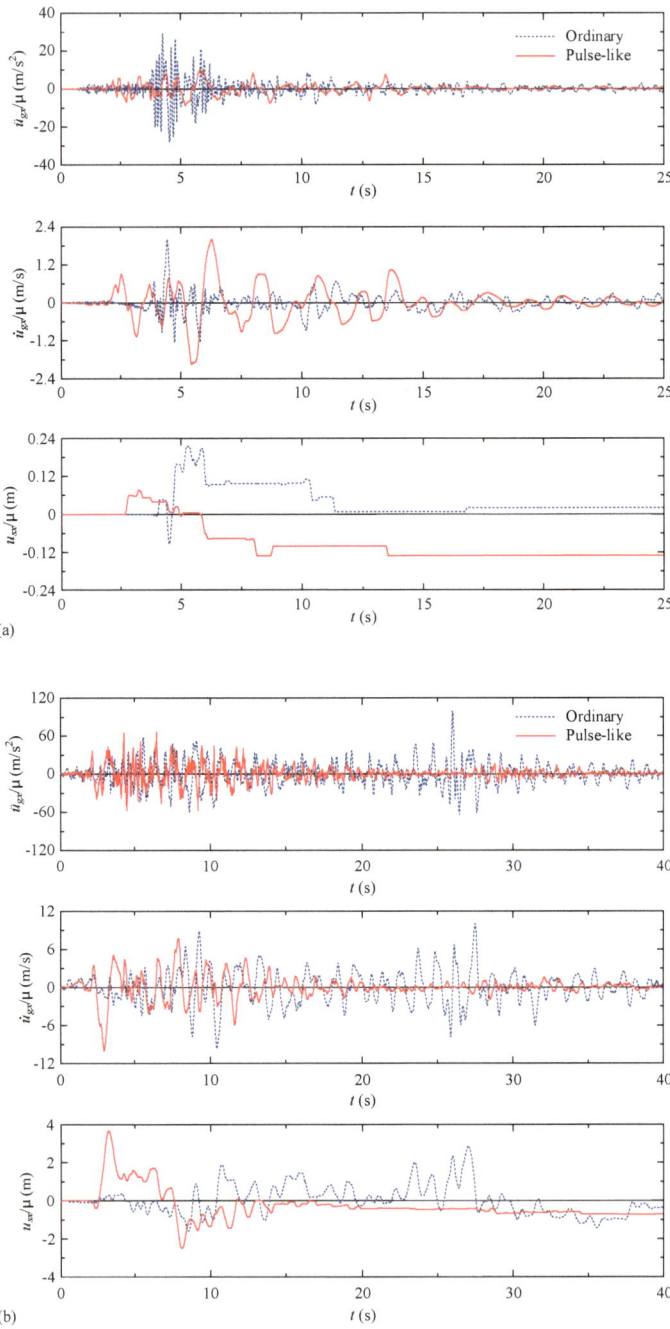

Fig. 6.20 Comparison of the time histories corresponding to the counted median of u_{sx0}/μ for the non-pulse-like and pulse-like ground motions ($T_x = T_y = 0.4$ s, $\alpha = 0.7$): **a** $v_{gx0}/\mu = 2$ m/s; **b** $v_{gx0}/\mu = 10$ m/s

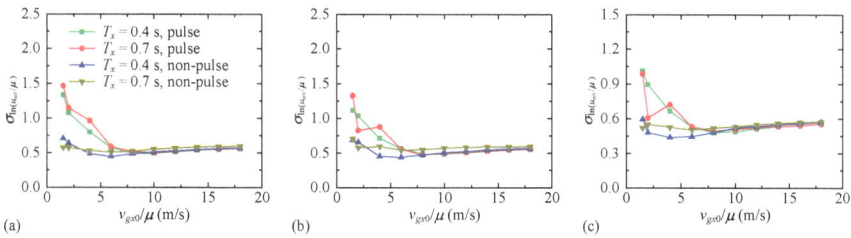

Fig. 6.21 Effects of near-fault pulses on $\sigma_{\ln(u_{sx0}/\mu)}$: **a** $\alpha = 0.5$; **b** $\alpha = 0.7$; **c** $\alpha = 0.9$

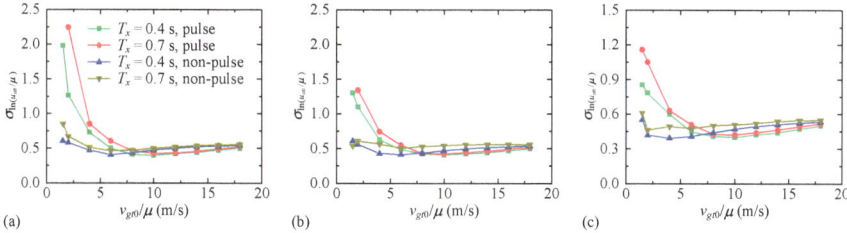

Fig. 6.22 Effects of near-fault pulses on $\sigma_{\ln(u_{st0}/\mu)}$: **a** $\alpha = 0.5$; **b** $\alpha = 0.7$; **c** $\alpha = 0.9$

6.4 Fragility Curves

From the investigations in the preceding section, we know that the influence of T_x and α on $(u_{sx0}/\mu)_m$ [and $(u_{st0}/\mu)_m$] is limited in comparison with that of v_{gx0}/μ (and v_{gt0}/μ). Therefore, in the design of SB structures, it is advisable to use the maximum values of $(u_{sx0}/\mu)_m$ and $(u_{st0}/\mu)_m$ conservatively for the common range of T_x. The maximum $(u_{sx0}/\mu)_m$ versus v_{gx0}/μ and maximum $(u_{st0}/\mu)_m$ versus v_{gt0}/μ curves are plotted in Fig. 6.23 for different values of α and for both the non-pulse-like and pulse-like ground motions. Since the curves corresponding to different values of α are very close to each other, for simplicity, equations for design can be developed solely based on the findings of $\alpha = 0.9$, which are generally larger than those of other values of α. It is found that a quadratic polynomial curve can well fit the relationship between the maximum $(u_{sx0}/\mu)_m$ and v_{gx0}/μ, as well as the maximum $(u_{st0}/\mu)_m$ and v_{gt0}/μ, and the obtained regression formulae are as follows:

(1) For the non-pulse-like ground motions,

$$(u_{sx0}/\mu)_m = 0.0052(v_{gx0}/\mu)^2 + 0.261(v_{gx0}/\mu) - 0.254 \geq 0 \qquad (6.12a)$$

$$(u_{st0}/\mu)_m = 0.0047(v_{gt0}/\mu)^2 + 0.283(v_{gt0}/\mu) - 0.308 \geq 0 \qquad (6.12b)$$

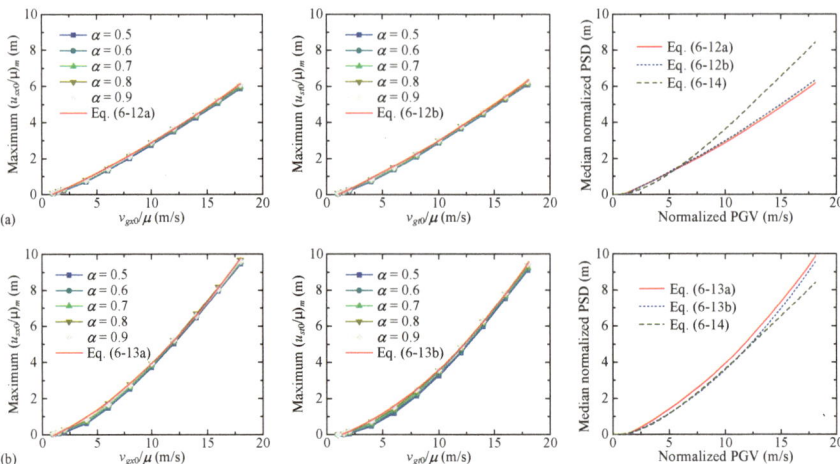

Fig. 6.23 Comparison of the design equations and numerical results for the relationships between the median normalized PSD and normalized PGV: **a** non-pulse-like ground motions; **b** pulse-like ground motions

(2) For the near-fault pulse-like ground motions,

$$(u_{sx0}/\mu)_m = 0.017(v_{gx0}/\mu)^2 + 0.257(v_{gx0}/\mu) - 0.341 \geq 0 \qquad (6.13a)$$

$$(u_{st0}/\mu)_m = 0.020(v_{gt0}/\mu)^2 + 0.190(v_{gt0}/\mu) - 0.282 \geq 0 \qquad (6.13b)$$

where $(u_{sx0}/\mu)_m$ and $(u_{st0}/\mu)_m$ are in m, and v_{gx0}/μ and v_{gt0}/μ are in m/s. According to the findings presented previously, replacing the subscript letter "x" with "y" enables the application of Eqs. (6.12a) and (6.13a) to the response in the y direction as well. As shown in Fig. 6.23, Eqs. (6.12) and (6.13) can well predict the corresponding relationships between the median normalized PSD and normalized PGV, and the coefficients of determination, R^2, of these equations are all larger than 0.99. Further comparison of the curves determined by Eqs. (6.12a) and (6.12b) [and Eqs. (6.13a) and (6.13b)], as presented in Fig. 6.23, indicates that the relationship between the median normalized PSD and normalized PGV in each principal direction is close to that with respect to the origin. Since the median normalized PSD versus normalized PGV curve corresponding to Eq. (6.12b) [and Eq. (6.13a)] is slightly above that corresponding to Eq. (6.12a) [and Eq. (6.13b)], Eqs. (6.12b) and (6.13a) can be used conservatively as unified equations for predicting the response in each principal direction as well as the maximum response over all the directions.

It is worth mentioning here that Ryan and Chopra (2004) also proposed a design equation for calculating the median peak displacements, $(u_{st0})_m$, of friction pendulum isolators:

$$(u_{st0})_m = \frac{4.36}{4\pi^2} T_b^{0.14} \eta^{(-0.99-0.20 \ln \eta)} \max(v_{gx0}, v_{gy0}) \tag{6.14}$$

where T_b is the isolation period, and η is defined as

$$\eta = \frac{\mu g}{\omega_d \max(v_{gx0}, v_{gy0})} \tag{6.15}$$

ω_d in Eq. (6.15) is the frequency marking the transition from the velocity-sensitive to the displacement-sensitive region of the median spectrum of the stronger horizontal ground-motion components. Note that in Eq. (6.14), the PSD with respect to the origin is taken as the response quantity of interest, while the PGV of the stronger component is taken as the ground motion IM, which sets it apart from the treatment in this study. The comparison of the median normalized PSD versus normalized PGV curve, determined by Eq. (6.14) [$\omega_d = 3.05$ is adopted, as done by Ryan and Chopra (2004), and T_b is taken as 10 s such that the corresponding radius of the FP isolator is sufficiently large to yield the same response as that of a flat sliding system], with those determined by Eqs. (6.12) and (6.13) in Fig. 6.23, is presented. As can be seen, the curve determined by Eq. (6.14) is close to those determined by Eq. (6.13), which is proposed for near-fault pulse-like ground motions. This is because the 20 ground motions used in the response history analyses conducted by Ryan and Chopra (2004) were from large-magnitude earthquakes and recorded at sites near fault ruptures, the characteristics of which are close to the near-fault pulse-like ground motions used in the present study.

As demonstrated in the preceding section, there does not exhibit a clear trend for the influence of the structural parameters, T_x, T_x/T_y, and α on the logarithmic standard deviations, $\sigma_{\ln(u_{sx0}/\mu)}$ and $\sigma_{\ln(u_{st0}/\mu)}$, and the values of $\sigma_{\ln(u_{sx0}/\mu)}$ [and $\sigma_{\ln(u_{st0}/\mu)}$] are generally between 0.4 and 0.6 except for some cases when v_{gx0}/μ (and v_{gt0}/μ) is small. Based on these results, adopting a constant value for $\sigma_{\ln(u_{sx0}/\mu)}$ [and $\sigma_{\ln(u_{st0}/\mu)}$] is reasonable in design. This value is taken as 0.55 for both $\sigma_{\ln(u_{sx0}/\mu)}$ and $\sigma_{\ln(u_{st0}/\mu)}$ here, which is approximately the average of all the results computed using the non-pulse-like ground motion records when v_{gx0}/μ (and $v_{gt0}/\mu) \geq 2$ m/s. As previously discussed, the dispersion for the pulse-like and non-pulse-like ground motions is expected to be the same; thus, the values of $\sigma_{\ln(u_{sx0}/\mu)}$ and $\sigma_{\ln(u_{st0}/\mu)}$ for the pulse-like ground motions are also taken as 0.55. The aforementioned dispersion is a result of the random nature of the ground motion, which belongs to the aleatory uncertainty. Other sources of variability are referred to as the epistemic uncertainty, which is related to the lack of knowledge about the real structural properties and modeling approximations. Simultaneous consideration of both types of uncertainty involves an elaborate Monte Carlo simulation with appropriate distribution functions for the structural properties, which requires considerable effort. For simplicity, an approximate method based on the assumption that the effects of aleatory and epistemic sources are independent (FEMA, 2009) is adopted here. Assuming a recommended epistemic dispersion of 0.35 for average modeling quality, as suggested by FEMA P-58-1 (FEMA, 2018), the total dispersion of the normalized PSD is $\sqrt{0.55^2 + 0.35^2} = 0.65$.

The fragility curve is an effective approach in assessing the seismic vulnerability of SB structures caused by excessive sliding, which presents the probabilities of exceeding a specified sliding displacement threshold at various levels of ground motion intensity. Since the normalized PSD (u_{sx0}/μ, u_{sy0}/μ, and u_{st0}/μ) at a given level of corresponding normalized PGV (v_{gx0}/μ, v_{gy0}/μ, and v_{gt0}/μ) follows the lognormal distribution, the probability, P_f, of exceeding the sliding displacement threshold, $u_{\lim}$, for given values of $PGV = pgv$ and $\mu = \mu_0$ can be computed by

$$
\begin{aligned}
P_f &= P(U_{s0} > u_{\lim} | PGV = pgv, \mu = \mu_0) \\
&= P[(U_{s0}/\mu_0) > (u_{\lim}/\mu_0) | (PGV/\mu) = (pgv/\mu_0)] \\
&= 1 - \Phi\left(\frac{\ln(u_{\lim}/\mu_0) - \ln(u_{s0}/\mu_0)_m}{\beta_{tot}} \right)
\end{aligned}
\tag{6.16}
$$

where U_{s0} represents the PSD of interest; Φ is the standard normal cumulative distribution function; the median normalized PSD, $(u_{s0}/\mu_0)_m$, is computed using Eq. (6.12) or Eq. (6.13); and the total dispersion, β_{tot}, is taken as 0.65, as discussed previously. Figure 6.24 shows the fragility curves for some typical values of μ and $u_{\lim}$ [Eqs. (6.12b) and (6.13a) were used in the computation], which clearly demonstrate the variation of P_f as the PGV increases and the effects of the primary parameters. As shown in Fig. 6.24, for small values of $u_{\lim}$ (e.g., $u_{\lim} = 0.1$ m) or large values of μ (e.g., $\mu = 0.4$), there is no significant difference between the fragility curves of the non-pulse-like and pulse-like ground motions; as $u_{\lim}$ increases or μ decreases, this difference becomes more significant and the SB structures subjected to pulse-like ground motions are more vulnerable in comparison with those subjected to non-pulse-like ground motions. This outcome agrees with the differences in the median normalized PSD and normalized PGV relationships depicted in Fig. 6.18 for both ground motion types.

6.5   Conclusions

This chapter presents a comprehensive study on the peak sliding displacements of SB structures subjected to three-component earthquake excitations. The PSDs in both the two main directions and with respect to the origin are taken into account. PGV is chosen as the ground motion IM because it exhibits a higher correlation with PSD compared to PGA and its attenuation relationship is conveniently accessible for design use.

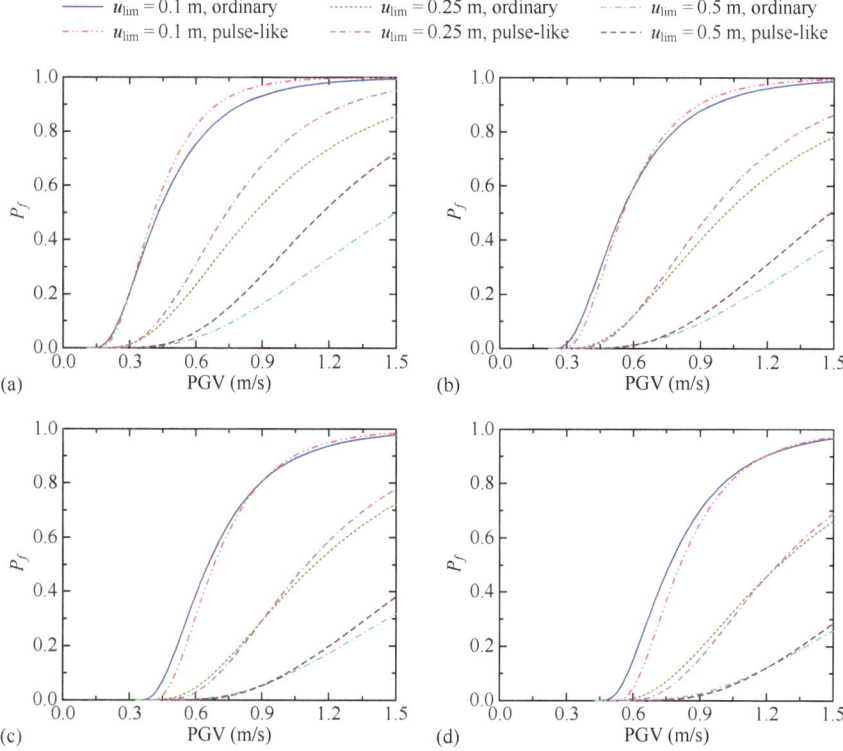

Fig. 6.24 Fragility curves: **a** $\mu = 0.1$; **b** $\mu = 0.2$; **c** $\mu = 0.3$; **d** $\mu = 0.4$

It is possible for an individual ground motion to exhibit significant differences in its PSDs between the two orthogonal horizontal directions, despite both directions having the same PGVs; however, on average, the relationship between the normalized PSD and normalized PGV is essentially identical for the two orthogonal directions. The effect of the vertical ground motion component on the PSD is negligible. The probability distributions of u_{sx0}/μ at a given level of v_{gx0}/μ and u_{st0}/μ at a given level of v_{gt0}/μ can be modeled by the lognormal distribution. The relationship between $(u_{sx0}/\mu)_m$ and v_{gx0}/μ and that between $(u_{st0}/\mu)_m$ and v_{gt0}/μ are close to each other. The influence of T_x, T_x/T_y, and α on $(u_{sx0}/\mu)_m$ and $(u_{st0}/\mu)_m$ is insignificant; thus, it is appropriate to conservatively use the maximum values of $(u_{sx0}/\mu)_m$ and $(u_{st0}/\mu)_m$ for the common ranges of T_x, T_y, and α in the design of SB structures. The lognormal standard deviation, $\sigma_{\ln(u_{sx0}/\mu)}$ [and $\sigma_{\ln(u_{st0}/\mu)}$], generally lies between 0.4 and 0.6 except for some cases when v_{gx0}/μ (and v_{gt0}/μ) is below 2.

When the normalized PGV is small, the values of $(u_{sx0}/\mu)_m$ [and $(u_{st0}/\mu)_m$] corresponding to the pulse-like records are close to those corresponding to the non-pulse-like records. When the normalized PGV exceeds a certain value (approximately

6–8 m/s), the value of $(u_{sx0}/\mu)_m$ [and $(u_{st0}/\mu)_m$] for the pulse-like records starts to exceed the corresponding value for the non-pulse-like records, and the difference increases monotonically as v_{gx0}/μ (and v_{gt0}/μ) increases. The difference in the value of $\sigma_{\ln(u_{sx0}/\mu)}$ [and $\sigma_{\ln(u_{st0}/\mu)}$] for the two types of ground motions is small.

References

Alhan, C., & Gavin, H. P. (2005). Reliability of base isolation for the protection of critical equipment from earthquake hazards. *Engineering Structures, 27*(9), 1435–1449.

Ang, A. H.-S., & Tang, W. H. (2006). *Probability concepts in engineering: Emphasis on applications to civil and environmental engineering.* Wiley.

ASCE. (2010). *Minimum design loads for buildings and other structures.* ASCE 7-10. ASCE.

Barbagallo, F., Hamashima, I., Hu, H., Kurata, M., & Nakashima, M. (2017). Base shear capping buildings with graphite-lubricated bases for collapse prevention in extreme earthquakes. *Earthquake Engineering & Structural Dynamics, 46*, 1003–1021.

Becker, T. C., & Mahin, S. A. (2013). Effect of support rotation on triple friction pendulum bearing behavior. *Earthquake Engineering & Structural Dynamics, 42*(12), 1731–1748.

Castaldo, P., Palazzo, B., & Vecchia, P. D. (2015). Seismic reliability of base-isolated structures with friction pendulum bearings. *Engineering Structures, 95*, 80–93.

Castaldo, P., & Tubaldi, E. (2015). Influence of FPS bearing properties on the seismic performance of base-isolated structures. *Earthquake Engineering & Structural Dynamics, 44*(15), 2817–2836.

Choi, B., & Tung, C. C. (2002). Estimating sliding displacement of an unanchored body subjected to earthquake excitation. *Earthquake Spectra, 18*(4), 601–613.

Chopra, A. K. (2001). *Dynamics of structures: Theory and applications to earthquake engineering.* Prentice Hall.

Chopra, A. K., & Chintanapakdee, C. (2001). Comparing response of SDF systems to near-fault and far-fault earthquake potions in the context of spectral regions. *Earthquake Engineering & Structural Dynamics, 30*(12), 1769–1789.

Chung, L. L., Kao, P. S., Yang, C. Y., Wu, L. Y., & Chen, H. M. (2013). Optimal frictional coefficient of structural isolation system. *Journal of Vibration and Control, 21*(3), 525–538.

Dolce, M., Cardone, D., & Croatto, F. (2005). Frictional behavior of steel-PTFE interfaces for seismic isolation. *Bulletin of Earthquake Engineering, 57*(3), 75–99.

Dolce, M., Cardone, D., & Ponzo, F. C. (2007). Shaking-table tests on reinforced concrete frames with different isolation systems. *Earthquake Engineering & Structural Dynamics, 36*(5), 573–596.

Eads, L., Miranda, E., & Lignos, D. G. (2015). Average spectral acceleration as an intensity measure for collapse risk assessment. *Earthquake Engineering & Structural Dynamics, 44*(12), 2057–2073.

Eröz, M., & Desroches, R. (2008). Bridge seismic response as a function of the friction pendulum system (FPS) modeling assumptions. *Engineering Structures, 30*(11), 3204–3212.

FEMA. (2009). *Quantification of building seismic performance factors.* FEMA P695. FEMA.

FEMA. (2018). *Seismic performance assessment of buildings. Volume 1: Methodology* (2nd ed.). FEMA P-58-1. FEMA.

Housner, G. W. (1941). Calculating the response of an oscillator to arbitrary ground motion. *Bulletin of the Seismological Society of America, 31*(2), 143–149.

Hu, H. S., Lin, F., Gao, Y. C., Guo, Z. X., & Wang, C. (2020). Maximum superstructure response of sliding-base structures under earthquake excitation. *Journal of Structural Engineering-ASCE, 146*(7), 04020131.

Hu, H. S., Lin, F., Xu, L. W., Gao, Y. C., & Guo, Z. X. (2022). Peak sliding displacements of sliding base structures subjected to earthquake excitation. *Journal of Structural Engineering-ASCE, 148*(3), 04021286.

Hu, H. S., & Nakashima, M. (2017). Responses of two-degree-of-freedom sliding base systems subjected to harmonic ground motions. *Journal of Structural Engineering-ASCE, 143*(2), 04016173.

Hutchings, I. M. (1992). *Tribology: Friction and wear of engineering materials.* Edward Arnold.

Hutchinson, T. C., & Chaudhuri, S. R. (2006). Simplified expression for seismic fragility estimation of sliding-dominated equipment and contents. *Earthquake Spectra, 22*(3), 709–732.

Iemura, H., Taghikhany, T., & Jain, S. K. (2007). Optimum design of resilient sliding isolation system for seismic protection of equipments. *Bulletin of Earthquake Engineering, 5*(1), 85–103.

Iura, M., Matsui, K., & Kosaka, I. (1992). Analytical expressions for three different modes in harmonic motion of sliding structures. *Earthquake Engineering & Structural Dynamics, 21*(9), 757–769.

Jampole, E. (2016). *High-friction sliding seismic isolation for enhanced performance of light frame structures during earthquakes* [Ph.D. dissertation]. Department of Civil and Environmental Engineering, Stanford University.

Jampole, E., Deierlein, G., Miranda, E., Fell, B., Swensen, S., & Acevedo, C. (2016). Full-scale dynamic testing of a sliding seismically isolated unibody house. *Earthquake Spectra, 32*(4), 2245–2270.

Jampole, E., Miranda, E., & Deierlein, G. G. (2018). Effective incremental ground velocity for estimating the peak sliding displacement of rigid structures to pulse-like earthquake ground motions. *Journal of Engineering Mechanics-ASCE, 144*(12), 04018113.

Jampole, E., Miranda, E., & Deierlein, G. G. (2020). Predicting earthquake-induced sliding displacements using effective incremental ground velocity. *Earthquake Spectra, 36*(1), 378–399.

Jangid, R. S. (1996a). Seismic response of sliding structures to bidirectional earthquake excitation. *Earthquake Engineering & Structural Dynamics, 25*(11), 1301–1306.

Jangid, R. S. (1996b). Seismic response of an asymmetric base isolated structure. *Computers & Structures, 60*(2), 261–267.

Jangid, R. S. (1997). Response of pure-friction sliding structures to bi-directional harmonic ground motion. *Engineering Structures, 19*(2), 97–104.

Jangid, R. S. (2005). Optimum friction pendulum system for near-fault motions. *Engineering Structures, 27*(3), 349–359.

Kelly, J. M. (1986). Aseismic base isolation: Review and bibliography. *Soil Dynamics and Earthquake Engineering, 5*(4), 202–216.

Kikuchi, M., & Aiken, I. D. (1997). An analytical hysteresis model for elastomeric seismic isolation bearings. *Earthquake Engineering & Structural Dynamics, 26*(2), 215–231.

Konstantinidis, D., & Makris, N. (2009). Experimental and analytical studies on the response of freestanding laboratory equipment to earthquake shaking. *Earthquake Engineering & Structural Dynamics, 38*(6), 827–848.

Konstantinidis, D., & Nikfar, F. (2015). Seismic response of sliding equipment and contents in base-isolated buildings subjected to broadband ground motions. *Earthquake Engineering & Structural Dynamics, 44*(6), 865–887.

Landi, L., Grazi, G., & Diotallevi, P. P. (2016). Comparison of different models for friction pendulum isolators in structures subjected to horizontal and vertical ground motions. *Soil Dynamics & Earthquake Engineering, 81*, 75–83.

Li, L. (1984). Base isolation measure for aseismic buildings in China. In *Proceedings of 8th World Conference on Earthquake Engineering* (pp. 791–798). San Francisco, USA.

Lin, S., Macrae, G. A., Dhakal, R. P., & Yeow, T. Z. (2015). Building contents sliding demands in elastically responding structures. *Engineering Structures, 86*(1), 182–191.

Lomiento, G., Bonessio, N., & Benzoni, G. (2013). Friction model for sliding bearings under seismic excitation. *Journal of Earthquake Engineering, 17*(8), 1162–1191.

Lopezgarcia, D., & Soong, T. T. (2003). Sliding fragility of block-type non-structural components. Part 1: Unrestrained components. *Earthquake Engineering & Structural Dynamics, 32*(1), 111–129.

Luco, N., & Cornell, C. A. (2007). Structure-specific scalar intensity measures for near-source and ordinary earthquake ground motions. *Earthquake Spectra, 23*(2), 357–392.

Maritz, J. S. (1995). *Distribution-free statistical methods*. Springer.

MathWorks Incorporation. (2014). *Curve fitting ToolboxTM user's guide (R2014a)*.

Mokha, A., Constantinou, M. C., Reinhorn, A. M., & Zayas, V. A. (1991). Experimental study of friction-pendulum isolation system. *Journal of Structural Engineering, 117*(4), 1201–1217.

Mostaghel, N., Hejazi, M., & Tanbakuchi, J. (1983). Response of sliding structures to harmonic support motion. *Earthquake Engineering & Structural Dynamics, 11*(3), 355–366.

Mostaghel, N., & Khodaverdian, M. (1987). Dynamics of resilient-friction base isolator (R-FBI). *Earthquake Engineering & Structural Dynamics, 15*(3), 379–390.

Mostaghel, N., & Tanbakuchi, J. (1983). Response of sliding structures to earthquake support motion. *Earthquake Engineering & Structural Dynamics, 11*(6), 729–748.

Nanda, R. P., Agarwal, P., & Shrikhande, M. (2012). Suitable friction sliding materials for base isolation of masonry buildings. *Shock and Vibration, 19*(6), 1327–1339.

Nanda, R. P., Shrikhande, M., & Agarwal, P. (2015). Low-cost base-isolation system for seismic protection of rural buildings. *Practice Periodical on Structural Design and Construction, 21*(1), 04015001.

Newmark, N. M. (1959). A method of computation for structural dynamics. *Journal of the Engineering Mechanics Division-ASCE, 85*, 67–94.

Newmark, N. M. (1965). Effects of earthquakes on dams and embankments. *Geotechnique, 15*(2), 139–160.

Qamaruddin, M., Arya, A. S., & Chandra, B. (1986a). Seismic response of brick buildings with sliding substructure. *Journal of Structural Engineering-ASCE, 112*(3), 558–572.

Qamaruddin, M., Rasheeduzzafar, Arya, A. S., & Chandra, B. (1986b). Seismic response of masonry buildings with sliding substructure. *Journal of Structural Engineering-ASCE, 112*(9), 2001–2011.

Roussis, P. C., & Constantinou, M. C. (2006). Experimental and analytical studies of structures seismically isolated with an uplift-restraining friction pendulum system. *Earthquake Engineering & Structural Dynamics, 35*(5), 595–611.

Ruiz-Garcia, J., & Miranda, E. (2007). Probabilistic estimation of maximum inelastic displacement demands for performance-based design. *Earthquake Engineering & Structural Dynamics, 36*(9), 1235–1254.

Ryan, K. L., & Chopra, A. K. (2004). Estimating the seismic displacement of friction pendulum isolators based on non-linear response history analysis. *Earthquake Engineering & Structural Dynamics, 33*(3), 359–373.

Shahi, S. K., & Baker, J. W. (2014). An efficient algorithm to identify strong-velocity pulses in multicomponent ground motions. *Bulletin of the Seismological Society of America, 104*(5), 2456–2466.

Shakib, H., & Fuladgar, A. (2003a). Response of pure-friction sliding structures to three components of earthquake excitation. *Computers & Structures, 81*(4), 189–196.

Shakib, H., & Fuladgar, A. (2003b). Effect of vertical component of earthquake on the response of pure-friction base-isolated asymmetric buildings. *Engineering Structures, 25*(14), 1841–1850.

Shao, Y., & Tung, C. C. (1999). Seismic response of unanchored bodies. *Earthquake Spectra, 15*(3), 523–536.

Shenton, H. W. (1996). Criteria for initiation of slide, rock, and slide-rock rigid-body modes. *Journal of Engineering Mechanics-ASCE, 122*(7), 690–693.

Skinner, R. I., Robinson, W. H., & McVerry, G. H. (1993). *An introduction to seismic isolation.* Wiley.

Taniguchi, T., & Miwa, T. (2007). A simple procedure to approximate slip displacement of free-standing rigid body subjected to earthquake motions. *Earthquake Engineering & Structural Dynamics, 36*(4), 481–501.

Tehrani, F. M., & Hasani, A. (1996). Behaviour of Iranian low rise buildings on sliding base to earthquake excitation. In *Proceedings of 11th World Conference on Earthquake Engineering,* Mexico City, Mexico, Paper 1433.

Tung, C. C. (2005). Criteria for initiation of motion of rigid bodies to base excitation. *Earthquake Engineering & Structural Dynamics, 34*(10), 1343–1350.

Vafai, A., Hamidi, M., & Ahmadi, G. (2001). Numerical modeling of MDOF structures with sliding supports using rigid-plastic link. *Earthquake Engineering & Structural Dynamics, 30*(1), 27–42.

Villaverde, R. (2009). *Fundamental concepts of earthquake engineering.* CRC Press.

Westermo, B., & Udwadia, F. (1983). Periodic response of a sliding oscillator system to harmonic excitation. *Earthquake Engineering & Structural Dynamics, 11*(1), 135–146.

Yamamoto, S., Kikuchi, M., Ueda, M., & Aiken, I. D. (2009). A mechanical model for elastomeric seismic isolation bearings including the influence of axial load. *Earthquake Engineering & Structural Dynamics, 38*(2), 157–180.

Yang, Y. B., Lee, T. Y., & Tsai, I. (1990). Response of multi-degree-of-freedom structures with sliding supports. *Earthquake Engineering & Structural Dynamics, 19*(5), 739–752.

Yegian, M. K., & Kadakal, U. (2004). Foundation isolation for seismic protection using a smooth synthetic liner. *Journal of Geotechnical and Geoenvironmental Engineering-ASCE, 130*(11), 1121–1130.

Zareian, F., & Krawinkler, H. (2007). Assessment of probability of collapse and design for collapse safety. *Earthquake Engineering & Structural Dynamics, 36*(13), 1901–1914.

Zayas, V. A., Low, S. S., & Mahin, S. A. (1987). *The FPS earthquake resisting system: Experimental report* [UCB/EERC-87/01]. Earthquake Engineering Research Center, University of California, Berkeley, CA.

Zhou, F. L. (1997). *Vibration control of engineering structures.* Seismological Press. [in Chinese].